RESEARCH METHODOLOGY AND SYSTEMS ANALYSIS

Encyclopaedia of Environmental Pollution and Awareness in 21st Century Series

RESEARCH METHODOLOGY AND SYSTEMS ANALYSIS

Edited by

V.K. Prabhakar

ANMOL PUBLICATIONS PVT. LTD.
NEW DELHI - 110 002 (INDIA)

ANMOL PUBLICATIONS PVT. LTD.
4374/4B, Ansari Road, Daryaganj
New Delhi - 110 002

Research Methodology and Systems Analysis

First Edition, 2001

ISBN 81-261-0929-7

PRINTED IN INDIA

Published by J.L. Kumar for Anmol Publications Pvt. Ltd., New Delhi - 110 002 and Printed at Mehra Offset Press, Delhi.

Contents

PREFACE

Today, the cry of 'pollution' is heard from all the nooks and corners of the world, and pollution has become a major threat to the very existence of mankind on this earth. It is the major challenge of our times.

The pollution of various resources has gone to such an extent that we are unable to breathe fresh air and drink fresh water. On one hand the advancements of science and technology have added to the human comforts by giving us automobiles, electrical appliances, supersonic jets, space crafts, better medicines, better chemicals to control harmful insects and other pests, etc., but, on the other hand, they have given us a very serious problem to face that is pollution. Pollution can be defined as an undesirable change in the physical, chemical, or biological characteristics of the air, water, or land that can harmfully affect health, survival, or activities of humans or other living organisms.

Pollution does not have to cause physical harm, but it may merely interfere with human activities. Chemicals spewed into the air or water from an industrial plant may be harmful to humans and other organisms living near the plant. If expensive pollution controls are required, the plant may be forced to shut down.

India faces serious problems of environmental pollution. Water pollution from untreated community and industrial wastes flowing into our rivers and streams, increasing use of fertilizers and pesticides, dumping of organic wastes etc., are issues that need to be tackled on the priority. Air pollution, long regarded as the bane of industrialised nations looms large over our country.

Here, we are presenting this *Encyclopaedia of Environmental Pollution and Awareness in Twenty-first Century* in fifty volumes with a view to make people conscious about the seriousness of pollution in the present time and its remedies.

Editor

1 SOCIAL AND ENVIRONMENTAL RESEARCH

Research of any kind is a careful investigation through search for new facts. Social research has been defined by P.V. Young in the following words—"We may define social research as the systematic method of discovering new facts or verifying old facts, their sequences, inter-relationships, causal explanations and the natural laws which govern them". On the basis of the above definition the following characteristics of social research may be deduced.

1. Social research deals with the social phenomena. It studies behaviour of human beings as members of society, and their feelings, responses, attitudes under different circumstances.
2. Social research is carried on both for discovering new facts and verification of the old ones. The object of every science, of course, is the discovery of new facts, new relationship, and new laws governing the phenomena. But constant verification of the old concepts is also needed, specially in case of dynamic sciences. Verification is needed because of two reasons. Firstly there may be an improvement in the technique of research and it is necessary to test the old concepts by this improved technique. Secondly the phenomena under study might have undergone a change and it may be required to test the validity

of old concepts in the changed circumstances. In Sociology a lot of research is being carried on for both purposes and has resulted in the discovery of new facts as well as modification of old concepts.

3. Social research tries to establish causal connection between various human activities. It is really very interesting to note whether various complex human activities, are being performed only at random without any sequence, law or system behind them. At the first casual look at varied human behaviour attitudes, moods and temperaments, the presence of any system may appear to be an impossibility, but a close and patient study of different cases, their scientific analysis and comparison and logical interpretation is bound to disclose the truth, that most of them are not merely haphazard as they appear but motivated by definite rules, perfect system and universal laws. The main purpose of social research is therefore to discover these laws, so that they may be used in the guided growth of human society.

Basic assumptions of social research

The application of scientific methods of sociological research is based upon certain assumptions, which may be classified as below.

1. Existence of cause and effect relationship — It has to be accepted as a basis of social research that there exists cause and effect relationship between various social activities. These causes always produce similar results and, therefore, if they are known they can be used effectively in checking the evils resulting from them. For example, social crime may be caused by poverty and frustration. In other words wherever there is poverty and frustration it will result in crime. Therefore, poverty and frustration have to be eliminated first in order to eliminate crimes.

2. Existence of sequence or law in social activities— Another assumption is that various social activities do not occur in a haphazard random way. There is some system some trend behind them. If this trend or system is located it is possible to predict the future course of social phenomena. For example, if it is proved that population of country increases by some specific rate, then population of the place in future years can safely be predicted.
3. Possibility of detached study — Even though man is an integral part of the society which he is investigating still it is possible for him to study it by keeping himself separate from the system. Hence his own feelings and emotions would not be reflected in the study. Even if such a study is a difficult task, it is possible.
4. Existence of ideal types — In a society, everyone is not entirely different from each other. People can be grouped into fairly, homogenous classes known as ideal type. The study made from this study can be made applicable to the whole society wherever it may exist.
5. Possibility of a representative sample — for research purpose, it is possible that instead of studying the whole population, a representative group called sample can be undertaken. The inferences drawn from this representative sample will apply to the whole population.

Objectives of research

The purpose of research is to discover answers to questions though the application of scientific procedures. The main aim of research is to find out the truth which is hidden and which has not been discovered as yet. Though each research study has its own specific purpose, we may think of research objectives as falling into a number of following broad groupings:

1. To gain familiarity with a phenomenon or to achieve new insights into it (studies with this object in view are termed as exploratory or formulative research studies);
2. To portray accurately the characteristics of a particular individual, situation or a group (studies with this object in view are known as descriptive research studies);
3. To determine the frequency with which something occurs or with which it is associated with something else (studies with this object in view are known as diagnostic research studies);
4. To test a hypothesis of a causal relationship between variables (such studies are know as hypothesis-testing research studies).

Motivation in research

What makes people to undertake research? This is a question of fundamental importance. The possible motives for doing research may be either one or more of the following:

1. Desire to get a research degree along with its consequential benefits;
2. Desire to face the challenge in solving the unsolved problems. i.e., concern over practical problems initiates research;
3. Desire to get intellectual joy of doing some creative work;
4. Desire to be of service to society;
5. Desire to get respectability.

However, this is not an exhaustive list of factors motivating people to undertake research studies. Many more factors such as directives of government, employment conditions, curiosity about new things, desire to understand causal relationships, social thinking and awakening, and the like may as well motivate (or at times compel) people to perform research operations.

Types of research

The basic types of research are as follows:

(i) *Descriptive vs. Analytical:* Descriptive research includes surveys and fact-finding enquiries of different kinds. The major purpose of descriptive research is description of the state of affairs as it exists at present. In social science and business research we quite often use the term Ex post facto research for descriptive research studies. The main characteristic of this method is that the researcher has no control over the variables; he can only report what has happened or what is happening. Most ex post facto research projects are used for descriptive studies in which the researcher seeks to measure such items as, for example, frequency of shopping, preferences of people, or similar data. Ex post facto studies also include attempts by researchers to discover causes even when they cannot control the variables. The methods of research utilized in descriptive research are survey methods of all kinds, including comparative and correlational methods. In analytical research, on the other hand, the researcher has to use facts or information already available, and analyze these to make a critical evaluation of the material.

(ii) *Applied vs. Fundamental:* Research can either be applied (or action) research or fundamental (or basic or pure) research. Applied research aims at finding a solution for an immediate problem facing a society or an industrial/business organisation, whereas fundamental research is mainly concerned with generalisations and with the formulation of a theory. "Gathering knowledge for knowledge's sake is termed `pure' or `basic' research."[4] Research concerning some natural phenomenon or relating to pure mathematics are examples of fundamental research. Similarly, research studies, concerning human behaviour carried on with a view to make generalisations about human behaviour carried on with a view to make generalisations about human behaviour, are also examples of fundamental research, but research aimed at certain conclusions (say, a solution) facing a concrete social or business problem is an example of applied research. Research to identify social, economic or political trends that may affect a particular institution or the copy research (research to find out whether certain communications will be read and understood) or the marketing research or evaluation research are examples of applied research. Thus, the central aim of applied research is to discover a solution for some pressing practical problem, whereas basic research is directed towards finding information that has a broad base of application and thus, adds to the already existing organized body of scientific knowledge.

(iii) *Quantitative vs. Qualitative:* Quantitative research is based on the measurement of quantity or amount. It is applicable to phenomena that can be expressed in terms of quality. Qualitative research, on the other hand, is concerned with qualitative phenomenon, i.e., phenomena relating to or involving quality or kind. For instance, when we are interested in investigating the reasons for human behaviour (i.e., why people think or do certain things), we quite often talk of 'Motivation research', an important type of qualitative research. This type of research aims at discovering the underlying motives and desires, using in depth interviews for the purpose. Other techniques of such research are word association tests, sentence completion tests story completion tests and similar other protective techniques. Attitude or opinion research i.e., research designed to find out how people feel or what they think about a particular subject or institution is also qualitative research. Qualitative research is specially important in the behavioural sciences where the aim is to discover the underlying motives of human behaviour. Through such research we can analyse the various factors which motivate people to behave in a particular manner or which make people like or dislike a particular thing. It may be stated, however, that to apply qualitative research in practice is relatively a difficult job and therefore, while doing such research, one should seek guidance from experimental psychologists.

(iv) *Conceptual vs. Empirical:* Conceptual research is that related to some abstract ideas(s) or theory. It is generally used by philosophers and thinkers to develop new concepts or to reinterpret existing ones. On the other hand, empirical research relies on experience or observation alone, often without due regard for system and theory. It is data-based research, coming up with conclusions which are capable of being verified by observation or experiment. We can also call it as experimental type of research. In such a research it is necessary to get at facts firsthand, at their source, and actively to go about doing certain things to stimulate the production of desired information. In such a research, the researcher must first provide himself with a working hypothesis or guess as to the probable results. He then works to get enough facts (data) to prove or disprove his hypothesis. He then sets up experimental designs which he thinks will manipulate the persons or the materials concerned so as to bring forth the desired information. such research is thus characterised by the experimenter's control over the variables under study and his deliberate manipulation of one of them to study its effects. Empirical research is appropriate when proof is sought that certain variables affect other variables in some way Evidence gathered through experiments or empirical studies is today considered to be the most powerful support possible for a given hypothesis.

(v) *Some Other Types of Research:* All other types of research are variations of one or more of the above stated approaches, based on either the purpose of research,or the time required to accomplish research, or the environment in which research is done, or on the basis of some other similar factor. From the point of view of time, we can think of research either as one-time research or longitudinal research. In the former case the research is confined to a single time-period, whereas in the latter case the research is carried on over several time-periods. Research can be field-setting research or laboratory research or simulation research, depending upon the environment in which it is to be carried out. Research can as well be understood as clinical or diagnostic research. Such research follow case-study methods or in depth approaches to reach the basic causal relations. Such studies usually go deep into the causes of things or events that interest us, using very small samples and very deep probing data gathering devices. The research may be exploratory or it may be formalized. The objective of exploratory research is the development of hypotheses rather than their testing, whereas formalized research studies are those with substantial structure and with specific hypotheses to be tested. Historical research is that which utilizes historical sources like documents, remains, etc. to study events or ideas of the past, including the philosophy of persons and groups at any remote point of time. Research can also be classified as conclusion-oriented and decison-oriented. While doing conclusion-oriented research, a researcher is free to pick up a problem,. redesign the enquiry as he proceeds and is prepared to conceptualize as he wishes. Decision-oriented research is always for the need of a decision maker and the researcher in this case is not free to embark upon research according to his own inclination. Operations research is an example of decision oriented research since it is a scientific method of providing executive departments with a quantitative basis for decisions regarding operations under their control.

Research approaches

The above description of the types of research brings to light the fact that there are two basic approaches to research, viz., quantitative approach and the qualitative approach. The former invovles the generation of data in quantitative formwhich can be subejcted to rigorous quantitative analysis in a formal and rigid fashion. This approach can be further sub-classified into inferential, experimental and simulation approaches to research. The purpose of inferential approach to research is to form a data base from which to infer characteristics or relationships of population. This usually means survey research where a sample of population is studied (questioned or observed) to determine its characteristics, and it is then inferred that the population has the same characteristics. Experimental approach is characterised by much greater control over the research environment and in this case some variable are manipulated to observe their effect on other variable. Simulation approach involves the construction of an artificial environment within which relevant information and data can be generated. This permits an observation of the dynamic behaviour of a system (or its sub-system) under controlled conditions.The term 'simulation' in the context of business and social sciences applications refers to "the operation of a numerical model that represents the structure of a dynamic process. Given the values of initial conditions, parameters and exogenous variables, a simulation is run to represent the behaviour of the process over time." Simulation approach can also be useful in building models for understanding future conditions.

Qualitative approach to research is concerned with subjective assessment of attitudes, opinions and behaviour. Research in such a situation is a function of researcher's insights and impressions. Such an approach to research generates results either in non-quantitative form or in the form which are not subjected to rigorous quantitative analysis. Generally, the techniques of focus groups interviews, projective techniques and depth interviews are used.

Sources of data

Data required for research can be made available from various sources. P.V. Young has classified them into two groups: (1) documentary and (2) field sources. Documentary sources include material already collected whether published or unpublished. Field resources may include living persons, scholars, scientists, research workers, leaders—people who have worked with the social group or have studied the problem. These persons may be consulted to find their opinion, experiences and ideas about the research problem. Among documentary sources the first place maybe given to the books written by different experts. These books are of two types. In the first category we include those books which deal with some specific theory e.g. Darvin's Origin of Species. In the second category are included such books as concern the description of the phenomena. The purpose of the author is a mere description of phenomena and not the criticism or support of any theory. Books of history,travel and autobiographies come under this category. Reports of official and unofficial survey and data published by various agencies also come under this group. The first type of books have theoretical importance while those of the second category have greater information value. Various sources of information are as follows.

(1) Documentary sources

1. *Books*-Different thinkers have described their ideas and theories in the books written by them. A study of all such books is essential for the researcher in order to know what has been said on the subject by people who conducted a similar research earlier. A study of all such work brings greater clarity and avoids duplication of effort. In the absence of such knowledge a researcher might be wasting his time upon a subject upon which much has already been said.
2. *Report of Surveys*—Various surveys pertaining to social problems are undertaken from time to time by government and non government agencies. Their reports provide a very useful material to the social research. Before using this data, the reliability of the institution, the time of research and their suitability for the present purpose must be kept in view.
3. *Memories*—Memories, autobiographies, life histories and letters at times provide useful information on many social subjects. Diaries maintained by

important people are sources of revealing information on many problems. Such documents have both social and psychological importance.

4. *Accounts of travels*—Accounts of travels are specially important in Anthropology. The travellers to Africa, and other less frequented parts of the world have given vivid description of native societies and their customs and these can provide useful material in Anthropological studies.
5. *Historical accounts*—A good deal material for social research can be available from history. Historical accounts is of two types: (i) the general history and (ii) the history of the particular phenomena. To quote "Like Computer himself the early British social scientists leaned heavily on historical or documentary evidence". If we accept social institution as product of evolution these historical accounts of advanced societies can help in throwing light upon the future or less developed ones.
6. *Official published data*—A large variety of data on social problems are regularly published by government in most of the countries. These data pertain to health, population, education, employment, crime etc. They can be used by social scientists for their own purpose. The data thus released are very much valued for the coverage, regularity and reliability of government as collecting agency both due to its resources and unbiased nature.
7. *Other unpublished record*—There is a lot of other unpublished material that lies buried in the official fields or personal records.

(2) Field sources

Besides the indirect sources of data described above there are following direct sources also.

1. *Direct observation-* According to this method the researcher himself collects necessary information by observing the phenomena. Observation may be carried on in the natural field setting or in the form of an experiment. If the scope of enquiry is small the researcher himself would undertake the work of observation, but if the nature of subject is such as to demand a wider field enquiry, help of the field workers may be taken. In the latter case standardisation of the system and proper supervision of the field workers is essential.
2. *Information from the informants-* If the phenomena is such as cannot be observed or if the presence of the observer is not possible, the only way to get information is through interviews. Information by means of interviews can be had either by contacting the informant directly or asking him to reply as set of questions by post. Direct interviews can also be either in the form of story-like narration or in the form of answers to specific questions.
3. *Information from witnesses-* This method is used by committees and commissions. Under this method the committee does not contact the persons about whom the enquiry is being conducted but those persons who are thought to possess sufficient knowledge of the subject. These persons, so interviewed, are known as witnesses. Thus, for example, in case of an enquiry about labour we may contact labour leaders, employees and labour experts.

Utility of social research

Social research is a time-consuming and expensive affair. A question may, therefore, be asked -what is the utility of all this effort, in a way this question seems to be some what uncalled for. It is not necessary that all our efforts should be motivated by some material benefit.

The chief material benefits of social research are as follows:

1. *Social control-* Planning is not confined to the field for Economics alone, it has an equal importance in the field of social organisation. We can no longer remain passive on-lookers and let the society take its own route whether towards good or evil. The leaders of society must guide its path for an orderly march towards pre-planned goals. A control over society is possible only when we have a complete knowledge of the organization and working of society and its various institutions; their inter-relationship and the motivations that guide human behaviour. All this can be achieved only through a scientific study of society.
2. *Social cohesion-* The study of society creates better understanding between different social groups, it reveals the underlying unity in the midst of apparent disparity, and thus helps to create the feeling of oneness, sympathy and understanding in place of racial prejudices. If such understanding is created a large part of national and international problems would be solved without much difficulty.
3. *Social welfare-* Social welfare or removal of social evils can be achieved through social research. Social research helps us to judge the magnitude of social evils and thus take necessary steps to remove them. A large number of legislative and reformative measures owe their origin to the reports of social surveys. Social research helps us to find out not only the magnitude but also the real causes of evils and thus helps us to devise ways to hit at the root of these evils.
4. *Social prediction-* Social research helps us to formulate social laws which show relationship between social

facts and their causes. Once the existence of causative factors and their magnitude is known we can predict the result. Although accurate prediction in majority of cases is not possible due to complexity of social phenomena, variety of causative factors and their unstability, yet certain broad trends can definitely be located. Statistical methods can successfully be used for such predictions. For example, we can predict with sufficient accuracy the population of a town or even a country by calculating the growth rate. The prediction gives a better control over social phenomena and helps in social planning.

5. *Social growth-* Social research helps in the growth lines. Every society is subjected to growth, the direction of which is influenced by its own structure, institutions, social values and motivations etc. But it is equally affected by our knowledge of our own society as well as other societies. Man is a rational being and gains by experience. The future path of social progress is, therefore, conditioned by our knowledge of ourselves and other people. Social research thus helps in guiding the trend of social growth on more proper lines and towards more cherished goals.
6. *Perfection of tool of research-* As social research progresses the tools of analysis and methods of research become more perfect with use. New tools are devised to give more accurate and precise results. As the tools of research become more perfect, the science of Sociology itself moves towards greater perfection.

Qualities of a good research worker

The success of a research work to a great extent depends upon the qualities of researcher. It is difficult to give hard and fast rules regarding the requisite qualifications and virtues of a researcher but certain broad qualities can be pointed out. The qualities are two fold:

(1) General qualities that are needed for any research.

(2) Specific qualities that are concerning the particular research.

(1) General qualities

The main general qualities necessary for the success in the research work are as follows:

1. *Scientific attitude*- The first essential quality of a successful research worker is that he must possess a 'scientific frame of mind' i.e. determination and ability to get at the naked facts and not to be influenced by mere appearances or by prevalent notions or by one's own wishes. 'Wolf' calls this attitude as sine qua none of all science. According to him, "he who is credulous enough to take things at their face value or is so lacking independence and initiative that he cannot break away from customary ideas or is so partial as to be influenced by his desires and wishes has not the making of a man of science".

2. *Imagination and insight*- A research worker must possess a high degree of imagination power. He must be able to prove deeper and deeper into the realm of abstract social phenomena and be able to catch the significance inherent in them. He should be able to grasp and visualize intangible aspects of society which forms a major part of the social phenomena.

3. *Perseverance*- The work of scientific research requires unlimited patience and inexhaustible perseverance. It is more so in case of social science where the phenomena has to be observed in the natural environment and the facility of guided research is

absent. Even in case of physical science sometimes it requires years of continuous labour to make an invention. The research worker, therefore, must be able to work patiently and continuously over long periods even when the possibility of near success is not there to encourage him.

4. *A quick grasping power-* The research worker should possess power to grasp the significance of things quickly. Most of the circumstances present themselves before the research worker with great suddenness, when he is not prepared to meet them. All such times he has to grasp their significance at once and react in proper time.

5. *Clarity of thinking-* The research worker should be very clear in his concepts. Clarity of concepts can also be attained by having a thorough and clear knowledge of the subject under study.

(2) Specific qualities

The Chief qualities required are as follows:

1. *Knowledge of the subject*—The research worker should have a complete knowledge of the subject under study. He should therefore, read all the texts on the matter, and form a clear-cut idea about the subject under study. The knowledge helps in preparing forms of questionnaire and schedule and getting proper information. He can enter into face to face discussion and remove any doubts arising in the minds of the people regarding the study.

2. *Knowledge of the technique of research*—The research worker should also possess the intimate knowledge of the technique that is to be applied to the problem. He should know its details and limitations. In the absence of such knowledge the research worker is likely to commit methodological errors and the deduction drawn by him may not be sound.

3. *Personal taste in the study-* The research worker should take personal taste in the study undertaken. This will always inspire him in work and keep high in times of difficulties. A forced work is often very monotonous and tiresome and may not yield very tangible results. But personal taste does not mean, that he should be party man. The research worker must be unprejudiced and free from all preconceptions.
4. *Familiarity about the informants-* It is desirable that the research worker should make himself familiar with the people whom he is studying. This is important from two points of view. Firstly, the familiarity of the research worker the people will help him to get intimate information. For example, refugees from West Pakistan would talk more freely to a person knowing their language, culture or living than to a stranger. The second advantage of familiarity is that in this way the researcher will be able to appreciate their views more properly than if he is completely unfamiliar with them. According to P.V. Young- "Interviewing can be carried on successfully only by one who is socially sensitive and responsive to the interview and is able to enter at least victoriously both into the interview process and into the social situations related to him by the respondent".
5. *Unbiased attitude-* The researcher should have no preconceptions about the subject under study. He should go to his research with absolutely clean slate. The following remarks of Mrs. Young about unbiased attitude are worth mentioning. "A scientist avoids personal and emotional interpretation of data. He is not a debtor taking issue with a side. He maintains an open mind and tests his finding and assumptions. He looks for the facts which would substantiate and give theory a new meaning and vitality".

Criteria of good research

Whatever may be the types of research works and studies, one thing that is important is that they all meet on the common ground of scientific method employed by them. One expects scientific research to satisfy the following criteria:

(1) The purpose of the research should be clearly defined and common concepts be used.

(2) The research procedure used should be described in sufficient detail to permit another researcher to repeat the research for further advancement, keeping the continuity of what has already been attained.

(3) The procedural design of the research should be carefully planned to yield results that are as objective as possible.

(4) The researcher should report with complete frankness, flaws in procedural design and estimate their effects upon the findings.

(5) The analysis of data should be sufficiently adequate to reveal its significance and the methods of analysis used should be appropriate. The validity and reliability of the data should be checked carefully.

(6) Conclusions should be confined to those justified by the data of the research and limited to those for which the data provide an adequate basis.

(7) Greater confidence in research is warranted if the researcher is experienced, has a good reputation in research and is a person of integrity.

In other words, we can state the qualities of a good research as under:

1. Good research is systematic: It means that research is structured with specified steps to be taken in a specified sequence in accordance with the well defined set of rules. Systematic characteristic of the

research does not rule out creative thinking but it certainly does reject the use of guessing and intuition in arriving at conclusions.

2. Good research is logical: This implies that research is guided by the rules of logical reasoning and the logical process of induction and deduction are of great value in carrying out research. Induction is the process of reasoning from a part to the whole whereas deduction is the process of reasoning from some premise to a conclusion which follows from that very premise. In fact, logical reasoning makes research more meaningful in the context of decision making.
3. Good research is empirical: It implies that research is related basically to one or more aspects of a real situation and deals with concrete data that provides a basis for external validity to research results.
4. Good research is replicable: This characteristic allows research results to be verified by replicating the study and thereby building a sound basis for decisions.

Problems encountered by researchers in India

Researchers in India, particularly those engaged in empirical research, are facing several problems. Some of the important problems are as follows:

1. The lack of a scientific training in the methodology of research is a great impediment for researchers in our country. There is paucity of competent researchers. Many researchers take a leap in the dark without knowing research methods. Most of the work, which goes in the name of research is not methodologically sound. Research to many researchers and even to their guides, is mostly a scissor and paste job without any insight shed on the collated materials. The consequence is obvious,

viz., the research results, quite often, do not reflect the reality or realities. Thus, a systematic study of research methodology is an urgent necessity. Before undertaking research projects, researchers should be well equipped with all the methodological aspects. As such, efforts should be made to provide short-duration intensive courses for meeting this requirement.

2. There is insufficient interaction between the university research departments on one side and business establishments, government departments and research institutions on the other side. A great deal of primary data of non-confidential nature remain untouched/untreated by the researchers for want of proper contacts. Efforts should be made to develop satisfactory liaison among all concerned for better and realistic researches. There is need for developing some mechanisms of a university-industry interaction programme so that academicians can get ideas from practitioners on what needs to be researched and practitioners can apply the research done by the academicians.
3. Most of the business units in our country do not have the confidence that the material supplied by them to researchers will not be misused and as such they are often reluctant in supplying the needed information to researchers. The concept of secrecy seems to be sacrosanct to business organisations in the country so much so that it proves an impermeable barrier to researchers. Thus, there is the need for generating the confidence that the information/data obtained from a business will not be misused.
4. Research studies overlapping one another are undertaken quite often for want of adequate information. This results in duplication and fritters

away resources. This problem can be solved by proper compilation and revision, at regular intervals, of a list of subjects on which and the places where the research is going on. Due attention should be given toward identification of research problems in various disciplines of applied science which are of immediate concern to the industries.

5. There does not exist a code of conduct for researchers and inter-university and inter-departmental rivalries are also quite common. Hence, there is need for developing a code of conduct for researchers which ,if adhered sincerely, can win over this problem.
6. Many researchers in our country also face the difficulty of adequate and timely secretarial assistance, including computerial assistance. This causes unnecessary delays in the completion of research studies. All possible efforts be made in this direction so that efficient secretarial assistance is made available to researchers and that too well in time. University Grants Commission must play a dynamic role in solving this difficulty.
7. Library management and functioning is not satisfactory at many places and much of the time and energy of researchers are spent in tracing out the books, journals, reports, etc., rather than in tracing out relevant material from them.
8. There is also the problem that many of our libraries are not able to get copies of old and new Acts/ Rules, reports and other government publications in time. This problem is felt more in libraries which are away in places from Delhi and/or the state capitals. Thus, efforts should be made for the regular and speedy supply of all governmental publications to reach our libraries.

9. There is also the difficulty of timely availability of published data from various government and other agencies doing this job in our country. Researcher also faces the problem on account of the fact that the published data vary quite significantly because of differences in coverage by the concerning agencies.
10. There may, at times, take place the problem of conceptualization and also problems relating to the process of data collection and related things.

Environmental research

Studies in environmental perception were recognised as an important concept in the explanation of man-environmental relationship in the 1960's. But in the writings of Sauer, Wright and Kirk the idea, that the people behave in the environment not on the basis of objective knowledge but in terms of subjective interpretation, has appeared much earlier in the late 1940's and early 1950's.

Since 1960 onwards, the concept has been taken up in various research projects, theoretically and empirically, at various levels of man-environmental relationship. Mental image and the process of image building have been the two aspects of the research. Of these two aspects, a large body of literature has grown up on mental image-its various facts of structural, appraisal and preferential sides along with its measurements and its applications. Structural side of image study has been analysed in a pioneer work by Lynch. Evaluation of environmental images has been dealt with by Burton and Kates and Saarien while preferential aspect of image was handled by Gould and White and Wolpret. But the aspects of development of perception process and its impact on human behaviour it is necessary to have a thorough understanding of mental images. In European tradition, image is handled first, while in School of Psychology human behaviour is observed first to analyse the process of image building. Whichever the method is, a comprehension of the link between the learning process, decision making and covert behaviour of human being needs to be clarified.

The study on environmental hazard conditions surely has given a different turn on man-environment relation from environmental determinism to environmental experience of subjective awareness of the real world and man's behaviour in that environment. The study of subjective image of environmental resources and management of resources has made it realistic and significant in explaining man's response to environment at different cultural situations.

The appraisal of the preferential aspects of environmental images has opened up the possibility of analysing the decision making process of human behaviour, say, migration. It is the subject image and relative comparison of places that matter most in location and interaction of human beings on space. Wolper's concept of "Place utility" for migration can be related to "environmental images".

The relevance of studied environmental images lies in the fact that , such a conceptual approach has cut across all areas of Human Geography. including man-environment relationship as well as man-man relationship at various scales of urban rural and regional studies. The majority of the studies are concerned with human behaviour and have given a new dimension to the theme of man-environment relationship. The following are only a few of the important studies on environmental perception at various levels of scale.

Neighboured study

In the study of neighbourhood, Lee showed that residents can conceive neighbourhood better than the whole city.

Lee asked residents of Cambridge to delineate their neighbourhood from their own experience. Some people had difficulty in drawing it, but the majority of them could do it. The concept of neighbourhood has been a puzzle to planners as well, because it is difficult to establish whether it is defined by social activities, economic facilities or physical morphology. Lee tried to measure involvement of people in the neighbourhood and the number of houses and shops it contained as proportion of the area and the number of shops within a circle of half a mile radius from house. Other variable included are social class, age, employment within the locality. He found that neighbourhood was perceived individually than by group.

Central area of a city

H.J. Klein tried to find out people's perception of central part of the city by verbal description, selection of location from a list of items. People have different opinion about the central business district. Residents though that the central area is closer to their home. In that study image is given more importance than behaviour.

In this present study, residents who have been interviewed, live in the central part of the city. In their opinion the central part of city is 'very good' in terms of low prices of commodities, high accessibility to shopping centres and work place, social interaction and transportation facility. Residents who have migrated to out of the central city also think that the core is 'good' from above points of view. But these do not think that the area is clean or has good facility for school and entertainment.

Perception of shopping centres

Dewn has carried out a study of perceptions of shopping centres in Bristol. Residents have a better idea of down town shopping centres than those of neighbourhood ones. Shopping centres were evaluated on the basis of price, design, ease of movement, visual appearance, range reputation, of goods service, shopping hours and atmosphere. Of these 'Price' was found to be most important factor. Down's study was mainly of methodological rather than conceptual one.

Image of the city

Lynch has done a pioneer work on image of study of Boston, Los Angeles and Jersey City. He put emphasis on legibility or visual form of attraction. He classified place into five types-paths, edes districts, noeds and landmarks. Lynch presented results for groups of residents rather than individuals. Results were obtained through different types of responses: street interviews, verbal descriptions and sketch maps. Lynch's study has stimulated a large amount of subsequent work. Lynch neglected the importance of the functions of places and values which residents attach to them.

Regions and nations

The concept of "resident desirability" was used by Gould and White of a study of school children in U.K., U.S.A. Ghana, Nigeria. If focuses on evaluation rather than description of places. The study of appraisive type can possibly be useful for prediction of human behaviour. But Gould did not make such an attempt. His idea was to find "mental maps" of environment of school children. He prepared maps with the help of component analysis found out the spatial image of students from various areas.

Environmental research-an example

A group of researcher undertook a study in the city of Ahmedabad. These researchers surveyed a group of residents of the city. The purpose and objective of the research was to assess a host of environmental issues, which are given below:

(a) Residents: Their characteristics housing, health, environmental quality in which they live and their action space.

(b) Residents' perception and sensitivity to various problems.

(c) Residents' image of their own neighbourhood as well as other parts of the city.

Here we give below, as illustration to help you, as to how the researchers proceeded with their research work scientifically.

Methodology and data collection

Considering the nature of environmental image, attempts have been made here to analyse image in qualitative and quantitative and in absolute and relative points of view. Residents' image of environment has attributed to group and individual socio-economic variables. Residents' image and behaviour has been compared between two groups in the same environment and also between outmigrants and residents of the core city.

The study is heavily based on field survey and questionnaire. The method using questionnaire is not free from defect. Apart from proper response, residents may not be able to reveal their images and assess the environmental quality through single interview. This study did have the scope of multiple entries of residents on a longitudinal basis. In spite of the inherent problem with field questionnaires, it is perhaps one of the important ways by which residents can be approached directly. The data gathered through this field questionnaire is termed 'Primary data'

Sampling technique of field questionnaire

Nearly 2,200 residents have been interviewed in the core

city. In addition, 240 residents who have migrated out of the core area have also been interviewed. The sample is of random stratified type by caste, religion, occupation and language, the size of the sample in each pol in a ward is proportionate to the population of residents in each ward, as there is not published data on residents by pol.

As there is no published information by caste or number of residents in pols, pols have been visited first and then categorised by "major" and "minor' caste or religion. If a community is seen to be more than 50 % in a pol, the pol has been named after that community under study. If such a community is seen more than in one pol, it has been taken up as a "major" community while other communities have been considered as "minor".

A pol is considered "mixed' if there is no majority of one community. There are, approximately 464 pols, with major and mixed communities, of which 114 pols have been taken for sampling (i.e. 25% of 464). There are a total of 41 pols, with "minor" caste of which 21 have been sampled; that makes a grand total of 135 pols to be sampled (i.e. 7% of the total 505 pols in the walled city). Finally, the selection of pols in a ward was decided in proportion to the distribution of particular community.

Secondary data

There is hardly any secondary data available on a spatial basis and on a micro level. Some of it are available in an aggregate level which does not cater to the needs of this study. There is very little data of important environmental factors like air, water and noise population and traffic congestion on special level. Data on all diseases and child mortality have been collected from the patient's registers of various hospitals onward basis-these are: density of population, density of houses.

Methodology and content of the field Survey

Initially a preliminary survey has been made on the basis of physical factors, e.g. cleanliness, level of physical factors, e.g. cleanliness, level of noise, density of traffic, population and quality or air. Then a pilot survey was conducted from three areas with three communities-one in the mill workers area, second one in the Thakkar Nagar of Sindle community and third one in Ambawadi area of the newly developed settlements on the western side of the river. Informal, long discussions with old residents of the walled city have given insight before understanding the actual survey work.

The evaluative image (attempted to be obtained via questionnaire) is, based on the following divisions:

(a) Assessment of environmental quality by physical aspects, housing condition, distribution of disease and crime and density of Population by qualitative and quantitative measures with the help of primary and secondary data;

(b) Residents' perception and sensitivity to various physical and social aspects of the environment,

(c) Investigator's perception to the same environmental issues which have been compared with that of residents,

(d) Residents image of their own neighbourhood in comparison with other areas of the city.

2 Methods of Data Collection

It is very important for a researcher to know very clearly about the various data collection methods and tools as well as their sources, because the most important issue in research especially in environ-mental research is to collect data. The accuracy of the research and report is based on the accuracy of data collected. The collection of accurate data depends on the fool proof system of data collection method adopted. So iWe are going to see here, the two most important methods of data collection viz. Interview and Observation. We shall also see two tools—namely Schedule and Questionnaire which aid in carrying out the above mentioned methods.

The two most important tools generally used in social surveys are schedule and questionnaire. Really speaking, the two forms are similar in nature and there is very slight difference between the two so far as construction is concerned. The only difference between the two is probably the fact that while the schedule has to be used in direct interview, or direct observation and is filled by the field worker himself, the questionnaire is generally mailed to the respondent, who fills it up and returns it to the sender. Thus, the main difference between the two is regarding the method of collection data through them. There is yet another form known as interview guide that contains only the topics or broad headings about which the questions are to be asked.

Interview guide is generally used in case of qualitative or in-depth interview.

Schedule

The schedule is the form containing some questions or blank which are to be filled by the workers after getting information from the informants. The schedule may, thus, contain two types of questions firstly those that are in form of direct questions and secondly those that are in form of a table. There are some kinds of information that can be procured only by putting a question, e.g. questions for eliciting the informants opinion, attitude preferences of his suggestions about some matter. There are others which may better be put in form of tables. Generally, most of them may be used interchangingly.

Purpose

The purpose of schedule is to provide a standardized tool for observation or interview in order to attain objectivity. By schedule every informant has to reply the same questions put in the same language and the field worker has no choice to get the desired reply by putting a different question or changing the language of the same question. The order of the question is also the same and thus the whole interview takes place under standardized conditions, and the data received is perfectly comparable. It has not been proved that responses of the people regarding the same matter differ, if the language of the question is even slightly different or even the place of question is changed. A perfectly standardized form is, therefore, needed for any objective study.

The other purpose of schedule is to act as memory tickler. In the absence of any schedule the field worker may put different number of questions to differen' people. He may forget to inquire about some importar aspects and then he would be required to go over the whole process again to collect that missing information too, the schedule keeps his memory refreshed, and keeps him reminded of the different aspects that are to be particularly observed.

Essentials of a good schedule

According to P.V. Young, there are two essential conditions of a good schedule. (1) Accurate communication and (2)Accurate response.

1. *Accurate Communication*—Accurate communication is achieved when respondents communication is achieved, when respondents understand the questions in the same sense which they are expected to convey. Thus the basis of accurate communications is proper wording of questions. The questions should be so worded that they may clearly carry the desired sense without any ambiguity. Various tools for achieving accuracy of communication have been discussed in the succeeding paragraphs.

2. *Accurate Response*—Accurate response is said to have been achieved when replies contain the information sought for. Thus information should be unbiased and true. The respondents may also co-operate in filling the schedule, by giving out correct information.

Following steps are generally taken to achieve the above two objectives :

1. The size and physical structure of the schedule should be attractive.
2. The questions should be clearly worded, easy to follow without any ambiguity or doubt.
3. Right type of questions should be asked.
4. Questions asked or information sought should be adequate and relevant to the enquiry, so that final generalizations may be based upon it.

5. The information should sought not be valid, it should be capable of being tabulated and if possible being subjected to statistical analysis.

Procedure for forming a schedule

While framing the schedule the first question to be asked is "what are the different aspects of the problem?" The problem under study should first of all be split up into various aspects. The determination of these aspects will depend upon clear understanding of the problem under study.

The second step is to be decided—"what information is necessary for a valid generalization on each aspect. For this purpose each aspect has again to be broken up into a number of sub parts. These sub parts should be quite exhaustive to give a full and complete picture of the aspect under study. Thus in the above illustration, care of the children may further be divided into care in respect of food, clothing, study, companionship, manners, health and character etc.

The third step is the framing of actual questions. This part deals with the form and working of the questions. More than one questions may be asked to get complete information about the particular aspect. Care should be taken to see that the questions convey the exact sense, are easily followed by the response and they will be willing to supply information without any hesitation, bias or distortion of facts. When information cannot be secured by direct questions, indirect questions may be resorted to. This part is the most vital part of the schedule and any error in it may invalidate the whole enquiry through biased, incorrect, incomplete or irrelevant information.

The fourth step is general layout of the schedule and arrangement of questions. When the questions have been given in a definite form the other problem is to arrange them in proper form. The layout of physical design of the schedule is very important, if it is well planned and attractive to look at, replies may be coming forth without hesitation.

The last step is testing the validity of schedule. After the schedule has been prepared, it has to be tested on a sample population to find out any discrepancies left out. Whatever may be the degree of precaution taken, some slips bound to be left out and can be located unless the schedule has been put into operation. Thus the schedule should first of all be prepared on trial basis, should be tried if any discrepancies in information sought. Woring of question, their sequence etc., should be amended in the light of the experience thus gained.

Type of questions

The different kind of questions that are asked in a schedule or questionnaire can be classified into the following types.

1. *Open-end questions*— These questions generally relate to opinion sought about any particular problems or suggestions invited about it. Naturally answers to these questions are long and varied. There is no prior classification or grouping into which replies are likely to be received. There is wide range of choice for the informant to give any reply. Not only the substance of the replies but also method of reporting may differ from person to person. Naturally open-end questions present a great difficulty in tabulation. Following are the examples of open-end questions.

1. What do you think are the main causes of present indiscipline among the students?
2. What suggestions do you offer to remove it?

2. *Structured questions*— When answer to a question has been preconceived and classified into possible groups it

is known as a structured question. Generally the answer to these questions may either be a number or some specific words. Thus, the possible replies to structured questions are already known and they are grouped into various classes. The tabulation of a structured question is thus very easy. There is greater standardization and objectivity in the replies. Following are the illustrations of structured question.

1. What is your age?
2. Are you educated/or uneducated?

3. *Dischotomous question*— When reply to a question is given in form of one of the two alternatives the question is called dischotomous. Generally, one of the answers is positive and the other negative, and two combined form the whole range of answers. Dischotomous questions are very good in classifying and tabulating. There is no danger of any possible alternative being left out. Given below is an illustration of dischotomous question.

Whether belongs to scheduled caste. Yes/No

4. *Multiple-choice questions*— They are also known as Cafeteria Questions. It is the opposite form of dischotomous questions. In these questions the reply is not confined to two alternatives only, but may be one among a number of possible replies. A great care has to be taken to see that all the alternatives have been included. Generally a few groups known as 'others' is included over and above the possible replies. Any omission of a group may seriously affect the nature of replies.

When the opinion about some matter is demanded take sell arrangement of the alternatives also affects the replies. A care should be taken to avoid any bias being caused in this way. One of the ways is to change the order in different sets. An illustration of the multiplier choice questions is given below.

1. What type of career do you propose to choose?
   - (a) Business or Industry
   - (b) Government Service
   - (c) Private Service
   - (d) Profession
   - (e) Vocation
   - (f) Any other (specify).

5. *Leading questions*— When reply to a question is suggested in a particular direction it is known as leading, suggestive or directive question. It is not necessary that reply to such a question may always be 'Yes' but the undecided cases are generally drawn towards it, thus creating a bias in the reply. Leading questions should be avoided as far as possible but when special emphasis has to be given to any particular aspect they may be included. Following is the illustration of a leading question.

1. Don't you think that education in municipal primary schools makes the children ill-mannered? 2. Isn't it more beneficial for the whole family that women, instead of doing service outside, should look after their family and children.

6. *Ambiguous questions*— When the language of the question is such that it may be interpreted in more than one ways, it is called an ambiguous question. Ambiguity in a schedule is the greatest defect and care should be taken in wording the questions in such a way that clear cut idea may be conveyed through them. Ambiguity may be caused in the following ways.

When the language or construction of the sentence is defective.

When some word or words having more than one interpretation are used. For examples,

What type of schooling did you have? 'schooling' here may mean type of school attended or type of education received.

7. *Ranking item question*— These questions are given to record the preference of the respondent. Like Multiple choice questions they also contain a number of alternatives. The only difference between the two is that whereas in case of the multiple choice questions only one of the given answers is to be selected, in case of ranking item questions order of preference is to be given for all. In this case also due care should be taken in giving proper sequence of the various alternatives in the schedule, as the order given in the schedule has great influence upon the respondents. An illustration of ranking question is given below :

1. What is your preference to following posts?
   - (a) Director of a mill.
   - (b) Indian Administrative Service
   - (c) Director of a Research Institute.
   - (d) Professor of a first rate University.
   - (e) A High Court pleader of good fame.
   - (f) A doctor of repute.

Questionnaire

We have seen that questionnaire is a set of questions used mainly in direct interview and to gather data from the respondent. We can put it this way that a questionnaire is simply a formalized schedule to obtain and record specified and relevant information with tolerable accuracy and correctness. In other words it directs the questioning process and promotes clear and proper recording.

A Questionnaire serves 5 functions :

1. Give the respondent clear comprehension of the question.
2. Induce the respondent to cooperate and to trust that answers will be treated confidentially.
3. Stimulate responses through greater introspection, plumbing of memory or reference to records.
4. Give instructions on what is wanted and the manner of responding.
5. Identify what needs to be known to classify and verify and interview.

Form of Questionnaire— The form of a questionnaire is similar to schedule as both of them serve the same purpose, but questionnaire has certain characteristics of its own because of the way in which data is collected. Following are the special points to be considered while collecting data with a questionnaire.

1. *Size of questionnaire*— The size of the questionnaire has to be smaller than the shedule.

2. *Appearance*— A much greater attention is required regarding the appearance of the questionnaire. In case of schedule the field worker is personally present to explain any doubt but in the case of questionnaire this has to be performed by the questionnaire itself. A decent paper, and printing, attractive lay out, good letter head creates a better impression upon respondent. In the absence of these qualities, the questionnaire may easily find its way to the waste paper basket. If the scheme has any backing of the government, the same must be mentioned in the accompanying letter. It generally has more salutary effect upon the response.

3. *Clarity*— The third important point that requires special attention in questionnaire is problem of clarity. In case of schedule the field worker is personally present to

give right interpretation to the questions and terms given in the schedule. In case of questionnaire this facility is absent. There is nobody to aid the respondent in interpreting a question. A great care has therefore, to be taken in drafting the questionnaire, the construction of questions and the language to be used. A thorough testing of questionnaire is required before it is finally sent to the respondents.

4. *Sequence of questions*— The sequence of questions is also of a greater importance in case of questionnaire than in case of schedule. A number of haphazard questions in the questionnaire fail to create the interest of the respondent. He begins to feel rather bored in filling it up and this may result either in careless half-hearted filling of the questionnaire, or in lack of response.

5. *Catching interest*— The greatest quality of a questionnaire is its interest-catching capacity. If it catches the interest of the respondent, it is sure to be carefully filled up and properly dispatched. If it fails to do so, the questionnaire is most probably doomed.

Problem of response questionnaires

In case of questionnaire the greatest problem is that of response. In an interview the field worker is personally present and very few people can refuse to grant interview at the face of the interviewer. In case of questionnaire, this facility is not present and the researcher has to rely on other means to get a good response.

Factors affecting response

A number of factors are responsible for varying degree of response. The chief of them are as follows :

1. *Special characteristics of the groups*— The response generally differs according to special characteristics of the respondent group. These special characteristics are sex, age, economic status, educational level etc. Starch made a

study of the influence of income group upon the degree of response. He has given the following findings.

Income group	% of response
0-999	2.6
1000-1999	2.5
2000-4999	14.1
5000-and above	9.0
All classes	7

This shows a clearly low degree of response among upper and lower income groups and even high among middle class people.

2. *Prestige of sponsoring groups*— If the institution carrying on research has a good reputation and credit, the percentage of responses is very high. People generally do not care to reply if the sponsoring authority is unknown, inconspicuous or a doubtful integrity, and the response is likely to be low.

3. *Importance of the problem under study*— If the problem under study is important and significant a higher response is expected. If it is only of a general nature the response is likely to be low. A higher response is expected specially, if the problem concerns the respondents directly. Studies by Reilly and Rtanton show that in marketing surveys persons who are users of products have higher response than non users. Reilly found that among the respondents who replied, 92% were the direct users only 80% non-users.

4. *Nature of questionnaire*— If the questionnaire is short and simple, its form is attractive and layout good, a higher response may be expected. If on the other hand,the questionnaire is clumsy, long or difficult the respondents are most likely to be indifferent and

apathetic. Therefore, in order to get a higher response, proper attention should be paid towards form and layout of the questionnaire.

5. *Nature of reaction*— According to Katz and Cantril "a very high percentage of respondents are those who have a strong feeling either in favour or against it." If the respondents are strongly in favour or against the problem, they are most likely to respond. If on the other hand, they do not possess any strong feelings and are more or less indifferent about the problem, they are expected not to reply.

6. *Size of questionnaire*— The size of the questionnaire also is an important factor in determining reason. Small size questionnaires generally have a better response than large size ones. But this advantage is available only when the size of the question is sufficiently small. After that small-size limit is crossed, any further increase in pages has no effect upon the response.

7. *Sequence of the questions*— According to Frazen, "the length of the questionnaire does not affect the response so much as the number of unrelated subjects dealt with." Thus, if the questions are not in one sequence and appear as broken subjects the response may be much lower than if they are in sequential order. Sletto's experiments also proves this fact. When complete questionnaire of 35 pages was sent it had a better response than 25 pages in complete questionnaire.

Interview

"Interview may be regarded as a systematic method by which a person enters more or less imaginatively into the life of a less imaginatively into the life of a comparative stranger." It is like a tape recorder in which the past incidences, feelings and reactions of the subject are played back to the living present to be listened by the interviewer with a scientific approach, always ready to find sequences of fundamental trials of human behaviour, underlying universal laws guiding and motivating human actions and reaction. The researcher tries to penetrate deeply in his imagination, into the circumstances being narrated by the subject and realize the full significance of the feelings being expressed by him.

The purpose of interview is two fold (1) Securing certain information from the subject, which is known only to himself and cannot be gathered from any source and (2) laboratory study of verbal behaviour under given circumstances. For the first purpose the interviewer tells the topic to the subject and he narrates the incidents of his life, his feelings and reactions pertaining to it.

The interviewer listens to these narratives, with cautions and alert,trying to find how far they fit in the hypothesis framed by him. For the second purpose the interest of the researcher is more centred in the attitude and expressions of the subject than the actual facts. Here he acts more as a social psychologist than as a sociologist. His attention is more focused at the reactions and changes in the moods and gestures of the subject while he is relating to a particular incident.

Kinds of Interviews

Interviews are generally of the following types :

1. Structured Interview

It is also known as controlled, guided or direct interview. In this kind of interview a complete schedule is used. The interviewer is asked to get the answers to those questions only. He generally does not add any thing from his own side. The language too is not changed. He can only interpret or amplify the statement wherever necessary.

2. Unstructured Interview

It is known as uncontrolled, unguided or undirected interview. No direct or predetermined questions are used in this type of interview. The field worker may be told certain broad topics upon which the information is to be collected. It is generally held in the form of free discussion or story type narrative. The subject is asked to narrate the incidents of his life, his own feeling and reactions and the researcher has to draw his own conclusions from it.

3. Focused Interview

Such an interviewee is generally used to study the social and psychological effects of mass communication e.g. the reactions of a film show or radio programme. Merton has given the following characteristics of a focused interview.

Interviewee is known to have been involved in a particular concrete situation, for example having seen a particular film or heard a broadcast. The situation under study is thus the one that has already been analyzed prior to interview. The field worker tries to focus his attention to the particular aspect of the problem, and tries to know his experiences, attitudes and emotional response regarding the concrete situation under study.

4. Repetitive Interview

Interview is repetitive in nature when it is desired to note the gradual influence of some social or psychological process. There are some social changes that have a far reaching influence upon the people and it is sometimes desired to know the effect of such factors in time sequence. Thus, for example, suppose a village is linked with some road connecting it with the city. Naturally,it will have its own influence upon the life of the people. The influence of course would not be sudden. There will be a gradual change in economic status, standard of living, attitudes and opinions, inter-relationships of the people. In order to study this influence in time sequence a study has to be conducted at regular intervals to mark the gradual change taking place.

Technique of interview

Conducting an interview is a technical matter and requires a thorough knowledge of the methods and the principles underlying it. No definite rules can be set for the purpose but some guiding principles can definitely be formulated and may prove useful if properly and timely used. Following are the different steps to be taken in conducting an interview.

1. Establishing contact

The first step in the interview naturally is to establish contact with the interviewee. If previous appointment has not been taken it may be a problem to get a contact with the interviewee and make him agreeable to grant an interview. As soon as the respondent appears at the door, the interviewer has to greet him and introduce himself. At this time he should produce the letter of authority, which must be on the printed form of the research institution, and contain general outline, importance and purpose of study. A letter is generally always better than oral introduction as it carried more weight, and produces an impressive effect. Moreover, while the interviewee is busy reading the letter, the research worker gets time to adjust himself to meet any questions from him.

2. Starting an interview

Once preliminaries have been completed and the respondent has agreed to grant an interview, the actual interview starts. Beginning of the interview is of great importance as it has its effect upon the entire information supplied. If the respondent is prejudiced at the very start he may not be giving free and frank replies so essential for interview. A beginning may be made from the general discussion of the problem. In order to warm up the respondent the interviewer should let him do most of the talking while he should himself listen to it attentively, guiding and directing the respondent about the subject matter wherever necessary.

3. Establishing rapport

According to Goode an Hatt: "a state of rapport exists between the interviewer and the respondent when the latter has accepted the research goals of the interviewer and activity seeks to help him in obtaining the necessary information." In the beginning of the interview every interviewee proceeds very cautiously giving only formal information, but if the researcher is tactful and expert in interview technique, he will soon create the friendly atmosphere; and gain the confidence of the respondent. No other device, but the researcher's own common sense, skill and personality can produce this state.

4. Recall

At times during the course of interview respondent is so full of emotional feelings that he drifts away from the main interview. Then when that narration has come to an end he generally lapses into silence. At such a time the researcher should wait sometime, waiting eagerly for him to continue again. When he has not started for sometime he must refresh his memory, by pointing out what he had been saying last.

5. Probe questions

There are occasions in the interview when the interviewer unknowingly or deliberately has side tracked some important aspect of the problem. At some times the researcher has to be very cautious to catch these slips. When the respondent has paused after finishing that part, the interviewer must draw his attention again to that fact feeling him. "You had been just telling me about. Would you tell me more about it; it is really so very interesting."

6. Encouragement

The respondent has to be encouraged from time to time during the course of interview when he narrates his feelings. Such complementary expressions as "what you have said is really very illuminating. I never had such an enlightening discussion. You really have a very unique approach to the problem. How strange, I myself had never thought of it from that angle etc." A great care should be taken that such remarks sound as true appreciation and not as mere flattery, otherwise they lose all their effect.

7. Guiding the interview

The interviewer has a very delicate duty to perform in case of narrative type interview. While he should enter fully into the spirit of the narrator with full appreciation and complete participation, he must be cautious not to get completely lost into it. He must maintain a detached view at the same time and skillfully guide the course of interview while it is drifting apart. It has generally been found that in the moment of emotional out-bursts the respondent is so absorbed that he drags the story to a point which is not of much importance to the study. At such times it is the duty of the interviewer to guide to the right path.

8. Recording

When the interview is free story type it is a problem to record it. In fact recording should be reduced to the minimum. When the interviewer begins to record the statement, the flow of the conversation is lost and the interview takes the form of questions and answers. Nobody can be expected to go on with his narration. While the other party is busy in something else, rather listening to him. Generally, when the interviewer begins to note, the respondent slows down his space and the flow is lost. Besides, when he is conscious that his statement is being recorded the free flow and objectivity is lost and undue caution overtakes him.

9. Closing the interview

How should the interview be closed. It should never be ended abruptly. After the respondent has narrated everything he wanted to, the tempo of his narration generally slows down. Occasionally he may relapse into silence, which continues to grow longer and longer as the interview gradually begins to come to an end. The interviewer should closely watch such a situation and give natural end to the discussion. It has generally been found that when rapport has gradually subsided the respondent is overtaken by a feeling of self remorse and awe. He begins to feel that he has given too much of his secrets to a stranger. Such a feeling although a criterion of a successful interview should be cautiously removed if possible.

It is generally useful to review the important points discussed before closing the interview. The interviewer may repeat the more important items and thus bring a natural close to the interview.

10. Report

When the interviewer has come home after interview he should immediately settle to write down what he has discussed. The writing of report should never be postponed at any cost. The various narratives, moods and expressions are vividly clear in the mind of the interviewer at this time and if sufficient time is allowed to pass, he may forget a great part of it. The report should also be as vivid and graphic as able, so that it may be able to create the same intensity of feelings at some later date.

Observation

Observation is probably the oldest method used by man in scientific investigation. His first knowledge of the universe around him begins with observation. The little child, when he opens his eyes in this world finds himself surrounded by strange and unknown things. His curiosity inspires and he begins to observe them. Gradually, he is able to understand their significance and use. So the process goes on.

Unaided sense observation was found to be crude and limited in power. So various kinds of tools were devised to increase their power, and make the observation more perfect. In modern times observation is not the only method of scientific research. Various other methods have been evolved but even now observation is one of the most important methods of research. In the physical sciences the observation is the only method for new experiment, although guided and controlled observation has replaced the ancient form of unaided visual observation.

Kinds of observation

Observation can be classified into various types according to the method used and the type of control exercised. Following are the chief types of observation.

1. Participant observation

The observation may be participant or non-participant. When the observer participates with the activities of the group under study, it is known as participant observation. Thus a participant observer makes himself part of the group under study. He need not necessarily carry out all the activities as carried out by other members of the group, but his presence as an active member of the group is necessary. Thus, he will be known as participant observer if he is present in some capacity other than as an observer viz. a historian, a correspondent etc. He freely mixes with social activities of the groups, their normal workings, functions and festivals, not as a total alien but as a full fledged member of the group, not as an outsider trying to have a critical study of what they do,but as a willing member trying to participate in what they do. Under participant observation, thus the purpose of critical study is concealed under the desire to learn folklores and mores, customs and traditions.

2. Non-Participant observation

When the observer does not actually participate in the activities of the group, but simply observes them from a distance, it is known as a non-participant observation. Purely non-participant observation is extremely difficult. We can not imagine a kind of relationship in which a person is always present but never participates. It is most likely to make both the observer and the group uncomfortable. A golden mean is therefore, sometimes, selected. The observer actively participates in some of the ordinary activities while remains only a distant observer in others. A non-participant observation is therefore, in practice only a quasi participant observation. It has been found easier to carry on both the roles than to attempt to disguise oneself completely.

3. Non-Controlled observation

Observation may be controlled or uncontrolled. When the observation is made in the natural surroundings, and the activities are performed in their usual course without being influenced or guided by any external force it is known as non-controlled observation. Thus in case of non-controlled observation the observer visits the place of occurrence of phenomena in order to observe e.g. study of slum conditions or folklores.

Non-controlled observation is generally not very reliable. We may be induced to generalize from stray incident, without properly verifying them. The observation itself may be biased and coloured by the views of the observer, because there is no check upon him. Various observers may observe the same thing differently and draw different conclusions. The greatest difficulty is that the observer may be so overpowered by uncontrolled and stray events that he may regard them to be absolutely true while they are far from being so.

4. Controlled observation

Because of the defects of non-controlled observation mentioned previously, controlled observation techniques have been developed. Such controlled observation afford greater precision and objectivity and can be repeatedly observed under identical conditions. The main purpose of a controlled observation is, thus, to check any bias due to faulty perception, inaccurate data and influence of outside factors on the particular incident. The control exercised in the observation is of two types.

(1) Control over the phenomena.

(2) Control over the observation.

Limitations of observation as a method of social research

Observation, although, widely used in social research has its own limitations. Not all occurrences are open to observation, can be observed when the observer is at hand, not all occurrences lend themselves to study by observational techniques. Thus the first drawback of observation is its uncontrolled nature. Observation of a phenomena has to be made where it actually takes place.

Most of the occurrences are quite uncertain in nature. Their place and time can not be predetermined. Consequently, a systematic and pre-planned observation is very difficult. There are three types of difficulties in the use of observation.

1. Some of the occurrences may not be open to observation

Incapacity for observation under this class may be caused due to refusal of the person concerned to allow the observer to which his activities. Most of the personal activities are thus not open to actual observation by an outsider. For example, if a study of the marital relations of the people is to be made no married couple would allow an outsider to observe their activities. It is only in case of social gatherings or functions of mass character that people can allow them to be observed.

2. Not all occurrences open to observation can be observed when observer is at hand

This type of incapacity is caused by uncertainty of the event. Many social events are most uncertain, so that the time and place of their occurrence can not be pre-planned. The event may take place when the observer is not present, on the other hand even when the observer is constantly present, the event may take place. Thus, for example, suppose we have to study the reaction of children when scolded or beaten by the parents. The study of such a reaction naturally depends upon existence of contingent situation viz. beating or scolding of the children by the parents. On principle the parents may have no objection to such a thing being observed. It is never certain when such an event will take place.

3. Not all occurrences lend themselves to study observational techniques

Most of the social phenomena to be studied are abstract in nature e.g. feelings, reactions etc. They are not open to our senses and hence cannot be subjected to study by observational techniques. For the study of such phenomena we have to adopt other techniques like case-study, interview etc.

The other limitation of the use of observation is caused by the lack of validity of the results. Even when the observational study has been made and data thus collected one problem remains to be solved. How far the data collected is valid to the problem under study. How far the information thus gathered is reliable.

Importance of observation in social research

Observation as method of social research has been used quite extensively everywhere. Following are the chief advantages of the method.

1. Simplest method

Observation is the simplest and the most non-technical method. Although scientific observation is not so easy, yet it is easier than other methods to follow by nature. A man is habituated to observe things that attract his attention. A little amount of training can make his observation perfect. Observation is also perfected with the theoretical knowledge of the phenomena which can be gained more easily.

2. Useful in framing hypothesis

Observation is one of the main sources of formulating hypothesis. The social researcher observes various activities of the people around him. Continuous observation even if causal and unplanned may reveal certain sequences and trends. These trends and sequences generally form the basis of new hypothesis. He must read and at the same time observe the activities of the people around him, see how far they agree with the theoretical principles and in what way do they differ.

3. Greater accuracy

Observation at times affords greater accuracy than other methods. In case of interview or case history the researcher has to depend on the information supplied by other people. But in case of observation the researcher need not depend upon information supplied by other people blind-folded. He can see if for himself. The data collected through observation is thus generally more valid and reliable, than data collected through observation, than data collected either through interview or case history.

4. More convincing results

The data collected through observation is more convincing than otherwise. When the information has been supplied by others, there is always some doubt in the mind of the people about its validity. Such data may, therefore, lack the force of conviction. In case of observation the researcher has observed thing for himself and no doubt about their correctness.

5. Test of validity

Observation is important in testing the validity of the hypothesis framed. Even if the researcher himself is fully convinced of the hypothesis, he has sometimes to prove its validity to others. This can be done through observation. He can let people observe the same things and this would convince them of the validity of the statement. Researcher himself also has to observe phenomena from time to time to see whether his generalizations hold good.

6. A common method for all sciences

Observation is a common method for all sciences. Interview is the method that is used specially in case of social sciences, but observation is common to all the sciences, and naturally has greater universality of practice. The rules of observation are the same for all sciences; whether physical or social. A common method is most commonly followed and accepted as a tool.

3 Analysis of Water Quality

Keeping in view the requirement of water in promoting industrial progress, maintaining public health, boosting up of agricultural production and almost in every walk of life methods for estimating altogether 30 different physico-chemical and biological parameters have hereinafter been briefly discussed so that the suitability of water for a variety of purposes be easily adjudged. For correct estimation of the parameters, proper methods of sampling and purity of the chemicals used in the analysis are equally important. Special care is needed in water sampling from different sources for the analysis of all parameters. The methods for sampling for different parameters are discussed below.

Sampling

The accurate assessment of the concentration of pollutants present in water primarily depends upon the samples drawn for the analysis. The samples must be representative in nature. The samples are generally of three types: (i) Grab samples are those which are drawn from a fixed point. Such samples reflect the existing physico-chemical condition in the system with respect to the time of the withdrawal of the samples. (ii) Composite samples are those which are collected from various parts of the same system for examination, over a fairly extended period. Such samples reveal the over all characteristics but do not reflect the anoxic conditions, existing in aquatic ecosystem, at the time of the sampling. (iii) Diurnal samples are those which are drawn in the grab manner from the fixed sampling sites at regular time intervals in a single day. Thus, the changing condition of the existing aquatic ecosystem, with respect to the varying intensity of solar radiation is known by this method. This type of sampling is more desirable in limnological studies where production is to be evaluated.

Frequencies and number of samples

The frequencies, duration and number of samples drawn, depend upon the purpose and the source from where the samplings are to be done. Supposing if diurnal variation in a pond or stream is intended to be measured, six hourly samples need to be drawn at least for complete 24 hours. Similarly, if potable water samples are to be drawn for bacteriological analysis from the public water supply establishment in a town of more than one lakh population, samples are to be drawn daily. If the number of persons ranging between 50,000 to 1,00,000 then sampling is required twice a week, whereas for population ranges between 20,000 to 50,000, samples of water need to be drawn once in a month. In case of small turbulent stream two samples are sufficient, one from upstream and another from downstream. In case of shallow lakes or ponds, water samples from 30 cm and 0.5 depth are required to be drawn for bacteriological and chemical analysis respectively. In case of deep lentic water bodies, several samples are drawn vertically.

The number of water samples to be drawn depends by and large on the area of the pond or lake to be studied. If these are 0.4 km^2 in area, only one samples is required, but if the area covered is more than 1.0 km^2 then a minimum of 3 water samples must be drawn. The quantity of sample to be drawn for plankton analysis depends on the approximate number of planktons present in the system. If the ponds are *oligotrophic*, 25 litres of water should be sieved through the suitable size of the plankton net. If the ponds happen to be *eutrophic*, a minimum of 5 litres of water samples are to be sieved through the net.

Site of water sampling

So far as the site of water sampling is concerned, too near or too distant sampling points should be avoided for chemical analysis. Stagnant zone of the surface water should be avoided for bacteriological samples. (The water samples should always be collected against the current and in case of persisting stagnation, an artificial current should be created while samples are being drawn for bacteriological studies.) On the contrary, the water samples should always be drawn from the calm zone of the rivers and ponds for identification and evaluation of the plankton.

Time interval between collection of water samples and their analysis

Some of the parameters such as the determination of pH, temperature, dissolved oxygen, free carbon dioxide and many other dissolved or evolving gaseous constituents must be analysed on the spot as soon as the samples are collected. For the rest of the chemical parameters, samples must be brought to the laboratory in well preserved condition, as early as possible, but in no case the time interval between the collection of the samples and the beginning of the analysis, should exceed more than 72 hours.

The samples collected for bacteriological studies must be subjected to analysis within 24 hours of their collection. The samples for the identification and evaluation of plankton must be preserved by adding 4 per cent form alin as preservative soon after the collection is over, preferably on the spot. Even then, their analysis is required to start at the earliest possible opportunity preferably within a month of collection of samples.

Methods of water sampling

The method of water sampling mainly depends on the nature of analyses to be carried out. Analysis may be broadly divided into physico-chemical, bacteriological and biological categories. The biological analysis includes the identification and quantitative estimation of physico and zoo-plankton, benthos and other biota generally found in the aquatic ecosystem.

(i) Water samples for chemical analysis

The water samples are collected from a depth of 0.5 m in case from open water bodies and in case of piped water supply system, from a tap, in thoroughly cleaned jar or natural glass containers of minimum 2.5 litres of capacity, provided with double cap device. The samples are collected upto the top, without leaving any space so as to prevent the premature release of dissolved gases during the transit period.

(ii) Water samples of bacteriological analysis

The water samples are collected from 30 cm depth in case of open water bodies and for a public hydrant after sterilising the tap for a few minutes with the help of a spirit lamp, prior to taking the samples from the same in properly sterilized neutral glass bottles of 120 ml. capacity. To each of these sampling bottles, 0.1 ml of 30 percent sodium thiosulphate solution is added prior to subjecting it to the process of sterilization.

The sample of water should not be collected upto the top. On the contrary, some space must be left for the bacteria to survive. The collected water samples are brought to the laboratory from the field in the packed sampling box and the analysis should be taken up positively within 24 hours of collection.

(iii) Water samples for plankton analysis

50 litres of water from the oligotrophic water bodies are sieved through plankton net, preferably of 0.1 mm mesh size. In case of eutrophic water bodies (plankton count more than 25,000/ml), no sieving is required. The algae so collected are immediately fixed by adding 1 to 2 ml of 4 per cent formalin solution as preservative. Alternative preservative has also been suggested for which the use of 1 ml of lugol's solution (prepared by dissolving 10 gKl, 5g sublimed iodine and 5 g sodium acetate in 100 ml distilled water) is indicated.

(iv) Water samples for benthic biota

The biota living in the bottom mud layer of the lakes, ponds or rivers, reflect the overall qualities of the water column standing over the stratum. The most conveniently used instrument of collecting benthos in the limnological studies is *Ekmann dredge* by which the mud samples from unit area are drawn. The samples of mud so collected are applied along with water to the combination of sieves, put together one over the other. The sieve of mesh size of 0.5 to 0.6 mm is kept above and another sieve of 0.2 m mesh size is kept below it. The biota separated are picked up with the help of forceps and put into a glass container of 50 ml capacity with wide mount, to which 10 per cent formalin solution is added as preservative. The samples of the biota so collected are further subjected to analysis.

Parameters analysed on the spot

Hydrogen-ion-concentration (pH)

The hydrogen ion concentration (pH) is as important as the total solid content, for the sewage and temperature for ecological studies as a whole. Waste water with an adverse concentration of hydrogen ion is difficult to treat by biological means and if the concentration is not altered before discharge, the waste water can alter the pH of natural waters. In domestic waste water (containing sewage and septic effluents), the pH value normally varies from 0 to 7.5 and therefore do not affect the receiving water body but an altogether different picture may arise in the case where acidic or alkaline wastes go into the streams.

Acidic wastes are corrosive to both metallic and concrete structures. Besides being toxic to the aquatic life, they react with the natural alkalinity of the water, there by increasing the carbonate hardness and thus rendering it unfit for future uses in laundry or boiler. On the other hand, water having pH value of more than 8 when discharged into the receiving stream, combines with free carbon dioxide and rather increases the alkalinity of the water. The optimum ranges of pH for aquatic life are 6.8 to 9.0.

Estimation

The pH value of clear water in the field is determined with the help of *Lovibond* pH comparator disc. Water samples are filled-up in the pH cell upto the marked calibration followed in quick succession with the addition of 3 to 5 drops of phenol red solution as indicator. After shaking the sample, the colour developed is matched in the disc by which the pH value is known. In case of coloured water sample or any coloured solution pH value is estimated by using electronic pH meter.

Dissolved Oxygen (DO)

The dissolved oxygen (DO) content shows the health and ability of the stream to purify itself through biochemical process. Too high content of dissolved oxygen is an indication of organic pollution, particularly when pollution is contributed by sewage out fall. When DO level drops below approximately 5 ppm, the more desirable species of fishes such as trout and brass level the area and only the coarse type of fishes predominate. Further it is reported that at an oxygen level of approximately 2 ppm, fishes disappear and the environment shifts towards anaerobic species. High depletion in oxygen content produces foul odour due to the anaerobic decomposition of organic waste leading to the formation of hydrogen sulphide.

Estimation

Do is estimated on the spot by adding 2 ml of manganese sulphate solution (prepared by dissolving 364 g $MnSO_4$, $4H_2O$, in 1 litre distilled water) to the freshly drawn water samples in glass bottles of 300 ml capacity. The content is mixed by swirling 2 ml of alkaline sodium iodide-azide (prepared by dissolving 700 g of sodium hydroxide, 135 g of sodium iodide and 10 g of sodium azide in 1 litre of distilled water), solution is then added. The whole content is thoroughly mixed and the resultant brown flock is allowed to settle for about 5 minutes. Thereafter, 2 ml of conc. H_2SO_4 is added whereby the settled flock gets dissolved and makes yellow solution on mixing 200 ml of the above solution is taken out in another cleaned neutral glass bottle and titrated against N/40 sodium thiosulphate solution using starch as indicator. The amount of titrant used gives the direct reading for DO in ppnn.

Temperature (°C)

2 ml $MnSO^4$ solute. +200 ml of water sample + 2ml of alkaline sodium iodide-azide mix brown flock + 2 ml H_2SO_4 (conc.) flock dissolved yellow solution titrated with sodium thinks starch. Temperature is one of the most important parameters for aquatic environment because

almost all the physical, chemical and biological properties are governed by it. For example, density, viscosity, surface tension and vapour pressure of water, more or less depend on the temperature profile of the system. It is the temperature which limits the saturation values of solids and gases that are dissolved in it. The rate of chemical reactions and other biological activity such as corrosion or incrustation, BOD, photosynthesis, growth and death of micro organisms are all dependent upon environmental temperature values. Indian climate provides almost an ideal range of solar temperature which attributes a great self purification strength in the stream. The rate of biochemical reaction is directly proportional to the environmental temperature in hottest summer months, the oxygen demand increases, leading to serious oxygen depletion problem in the systems concerned.

Estimation

The temperature of water measured by using a thermometer, calibrated in ^{0}C scale. For measuring the temperature of open water body, the thermometer is dipped into the water and the reading is taken. Thermometer should not be taken out from the water body for reading the temperature. If the thermometer of ^{0}C scale is not available, then one of ^{0}F may be used.

Turbidity/transparency

Turbidity of water is responsible for the light to be scattered or absorbed rather than straight transmission through the sample. *It is the size, shape and refractive index of the suspended particulates rather than the total concentration of the latter present in the water samples that are responsible for turbidity.*

Turbidity decreases light penetration, limits production of phytoplankton which consequently leads to decrease in photosynthetic activity and depletion of oxygen content. Clear ponds with less than 25 ppm turbidity may have 12.8 times more plankton and 5.5 times more fish production than ponds with a turbidity exceeding 100 ppm. High turbidity closes the door and impairs the lungs by completely blanketing the upper surface of water body. Since the surface not only acts as window through which it receives radiant energy but it is also like lungs through which it takes in oxygen from the atmosphere and gives out carbon dioxide and other dissolved gases that are generated within the water by microbes. The turbidity and the transparency values are interrelated, just like that of total dissolved solid and conductivity values. The transparency values are expressed in cm or mm and turbidity in ppm. The values of transparency and their corresponding values for turbidity (in ppm) are shown in the table:

Table 1

Relation between transparency (mm) and turbidity (ppm)

Transparency (mm)	Turbidity (ppm)	Transparency (mm)	Turbidity (ppm)
20	1000	32	600
22	900	34	450
24	800	100	100
26	700	328	25
28	600	390	20
30	550	490	10

Odour

Odour in water is caused by *volatile substances* associated with organic matter, living organisms principally algae and related organisms and gases. Offensive odour causes poor appetite for good food, lowers down the water consumption, impairs general quality and causes nausea and vomiting. Odour can result in marked decline in tax revenues, pay rolls and sales. World Health Organisation (WHO 1961) has recommended that water intended for drinking, bathing and recreational purposes should be completely free from any odour.

Colour

The dissolved or colloidal substances of vegetable origin which generally consists of tannins, glucoside as well as iron and other substances are responsible for the colour present in water. The colour of sewage reflects its strength and condition. Fresh sewage is grey where as septic sewage is black in colour. The colour of river water is reduced by storage or aging of the water and by bleaching action of sunlight.

Estimation

1 g of cobalt chloride along with 1.245 g of potassium chloroplatinate are dissolved in distilled water and the solution is then diluted to 1 litre when the same has a colour of 500 hazen unit. Maximum permissible colour for drinking purposes is 20 ppm on platinum cobalt scale. Colour is harmless but it cannot be allowed more than the maximum tolerance limit fixed as the consumers feel repellant.

Physio-chemical parameters

Total dissolved solids (TDS)

TDS contents of water and waste water are defined as residue left upon evaporation at 103 ^{0}C to 105^{0}C. It is an aggregated amount of the entire floating, suspended, suitable and dissolved solids present in the water sample. Further, in all the three above mentioned physical stages, the solids may be organic or inorganic in nature, depending upon volatility of the substances. The organic fraction is driven off as gas on ignition at 600^{0}C and the inorganic fraction remains as ash. Thus, the term volatile and non-volatile solids refer respectively to organic and inorganic contents of the water sample. Organic constituents of total solids are responsible for taste, colour, odour, gas and biological problems, whereas most of inorganic materials being non-biodegradable do not cause such environmental pollution problems, although a few of them, in traces, produce acute toxicological problems for aquatic dwellers.

Inorganic mineral contents, present in the water, conduct electric current which is measured by specific conductivity tests. An approximate idea of total dissolved solids (TDS) in natural water can be obtained by multiplying the specific conductance at 25^{0}C by a factor of 0.55 to 0.75 with an average of 0.65. TDS is ppm can also be found out by adding the sum of the determined constituents in ppm except that the bicarbonate concentration should be multiplied by a factor of 0.49. Osmotic pressure in atoms is known on multiplying the conductivity value of m hos/cm by 0.00036. Studies carried out on the productivity of fishes with respect to conductivity values suggest that electric conductivity above 400 m hos/cm do not limit production. On the other hand, the fish production does not increase proportionately with conductivity.

Alkalinity

It is an anionic phenomenon. All anions such as CO^{-2}_2 (carbonate), HCO^{-3} bicarbonate), OH (hydroxyl), PO_4^{-3} (phosphate), SiO_4- (Silicate) etc. contribute alkalinity to water. The number of mili-equivalents of acid used in the titration to combine all the hydroxyl ions, is known as total alkalinity. The water supplied with less than 103 ppm of alkalinity as $CaCO_3$ is ideal for domestic use. The ratio of alkalinity to that of alkaline earth metals is a good parameter for determining the suitablility of irrigation water.

Estimation

To 100 ml of water sample, 2 drops of 0.1 per cent phenalphthalein (P) solution, as indicator is added and titrated against N/50 H_2SO_4 solution, until the pink colour disappears. Thereafter, another 2 drops of 0.1 percent methyl orange solution as indicator is added to the same sample of water. Titration against N/50 H_2 SO_4 is continued until the yellow colour changes to pink. The titrant consumed in ml give the value for total alkalinity (TA). The calculation for different types of alkalinity involves some mathematical interplay which is as such:

(i) When P=0, TA is HCO_3 alkalinity as $CaCO_3$.

(ii) When P = TA, TA is OH alkalinity as $CaCO_3$.

(iii) When P = 1/2 TA, TA is CO_3 alkalinity as $CaCO_3$ and equal to 2 P. But when P is either less or more than half of total alkalinity (TA), then altogether a different situation aries. More than one type of alkalinity exists in the same sample.

(a) When P<1/2 TA, then CO_3 will be 2 P and HCO_3 will be T-2P.

(b) When P>1/2 TA, then 2 P-T will be for OH and 2 (1/2-P) will be for carbonate type of alkalinity.

Hardness

Hardness is caused due to bi-valent cations such as Ca^{++}, Sr^{++} etc. Higher cations also contribute hardness to a lesser degree but mono-valent cations never produce hardness. Thus, sodium compounds do not cause hardness from the agricultural, limnological and industrial point of view. The estimation of hardness is of vital significance but has a very limited scope in respect of environmental considerations. Water should have zero hardness specially for boilers operating at high pressure. Limnologicaly, moderately hard water is desirable.

Estimation

To 100 ml of water sample, 2 ml of buffer solution (prepared by dissolving 16.9 g NH_4 Cl in 143 ml of concentrated ammonium hydroxide solution) is added to bring the pH value to 9.0. Then one small spoon (say 0.209) of Erichrome black-T is added as indicator which forms wine red colour. The whole content is stirred and titrated against 0.01 M solution of EDTA (ethylene diamine tetra acetic acid), till the wine red colour changes to blue. To avoid the possibility of interference by trace elements, if suspected to be present in the water, one ml of 5 per cent sodium sulphide (Na_2 S) solution is added as inhibitor, prior to the procedure for titration. The titrant consumed in ml on multiplying by factor 10 gives the value for the total hardness in ppm as $CaCO_3$.

Relation between hardness and alkalinity

When the total alkalinity value is higher than the value for total hardness, then latter indicates the carbonate type of hardness. If and when the total hardness is higher than the total alkalinity, then the value for total alkalinity, is for carbonate hardness and the difference of the two shows the value for non carbonate hardness

Calcium

Calcium is one of the alkaline earth metals, the others being magnesium, barium, strontium etc. It is widely distributed on the earth's crust and is present nearly in all waters. In presence of CO_2, Calcium bicarbonate can normally be dissolved upto 20 ppm (as Ca) at atmospheric pressure and upto 100 ppm at higher pressure. The water coming from limestone zone may contain greater concentration of calcium. As soon as CO_2 is released, calcium carbonate (which can be retained in water upto 5 ppm as calcium only) is precipitated out. Lakes have been classified as poor, medium and rich, depending upon their calcium content, being less than 10 ppm, between 10 to 25 ppm and more than 25 ppm respectively. Calcium is not known to indicate or produce any hazardous effect on human health. Concentration upto 1800 ppm has been reported as not to impart any physiological reaction on human being. Moderate content of calcium in water is rather desirable because the toxicity of various substances i.e. Pb, Cd, Hg etc. have been observed to be neutralised. Undoubtedly, high concentration of calcium content in water is not desirable at all for laundering and for use in boiler. Calcium, if present with sulphate, inhibits malt fermentation and with chloride, inhibits growth of yeast in distillery industry.

Estimation

2 ml of 1 N NaOH solution is added to 100 ml of water sample, to raise the pH above 10. At such high pH, magnesium is precipitated out as Mg $(OH)_2$ but Ca remains in solution. It is then titrated against 0.01 M solution of EDTA, by using muroxide as indicator. The titration is continued till the solution becomes distinctly pink in colour. The titrant consumed in ml, on multiplication by factor 4, gives the concentration of calcium as Ca (in ppm). Calcium as $CaCO_3$ is obtained in ppm by multiplying the titrant value in ml by 10.

Magnesium

Its concentration is known by direct mathematical calculation (in ppm) as Mg. The formula used is as such:

Magnesium as Mg in ppm = 0.24 X (Total hardness as $CaCO_3$- Calcium as Ca)

Magnesium has ten times the solubility of calcium and being bivalent, it too produces hardness. It is absolutely essential for the chlorophyll bearing bacteria, algae and płants. It is second to calcium in hard water, whereas sodium (Na) becomes second to calcium in shoft water. Na is metabolized only by blue green algae but potassium (K) alone with magnesium is a necessary requirement for all algae. Under low Mg and K conditions, growth and photosynthesis of algae are poor and respiration is high.

Chloride

The sudden increase in the concentration of chloride in surface water bodies at abnormally high concentration was previously used as index of pollution through contamination by fecal matters. Human excreta contains chloride equal to the chloride consumed with food and water. Normal human body discharges from 8 to 15 g(av 9 g of chloride) a day. Since chloride ions are non-biodegradable, these are not removed from the wastes even after subjecting it to the secondary treatment process. However, it is harmless upto 1500 ppm concentration but produces a salty test at 250-500 pmm level.

Estimation

100 ml of water sample is taken in white porcelain basin to which 1 ml of 5% potassium chromate solution, as indicator, is added an titrated against standard silver nitrate solution (prepared by dissolving 4.79 g of $AgNO_3$ in one litre). The titrant consumed in ml on multiplying by 10, gives the direct reading for chloride content of water in ppm.

Fluoride

Flouride concentration is water both at levels less than 1.0 ppm and more than 1.5 ppm, are harmful for health. According to UNESCO (1963) and California State Water Pollution Control Board (1952), water containing more than 1.0 to 1.5 ppm of fluoride can cause mottled enamel in children and an excess of fluoride may eventually cause endemic cumulative flurosis with resultant skeletal damage in both children and adults. If the flouride concentration in drinking water is less than 0.5 ppm, a high incidence of dental carries is likely to occur. Recent researches by Geological Survey of India have led to the discovery of a natural mineral setpentine which can absorb large quantities of fluorides from water and can act as a defluoriding agent with considerable success in the treatment of fluoresces.

Estimation

50 ml of water sample along with control solutions are taken in different nesselar tubes and put on the porcelain based wooden stand. To each of the tubes, 2 ml of zirconium alizarine solution is added. 90 minutes after the addition of reagent, the developed colour is matched with the control solution and accordingly calculated. The colour changes from pink to yellow depending on the concentration of the flouride present in water samples.

Iodine

About 50 mg of iodine is present in a human body weighing about 70 kg. Probably all the cells of the body contain iodine, but 70 to 80 percent of total iodine content of the body is concentrated in the thyroid gland alone. When the dietary iodine intake falls below 10 mg/day throxine synthesis becomes, inadequate leading to endemic goiter. Persons suffering from goiter become mentally dull, lethargic and lack motor coordination capacity. Tetraglycine hydroperiodide or galbaline tablets or iodised salt are available in the market for compensating the deficiency of iodine in water and diet.

Estimation

The method for iodine determination begins with the

addition of 1 ml of 1 percent of arsenious oxide solution to each of the 10 ml of collected water samples. After thoroughly mixing the content, 1 ml of ceric ammonium sulphate solution (prepared by dissolving 13.38 g of 44 ml conc. H_2SO_4 and diluted to one litre) is added. Within 15 minutes, I ml of ferrous ammonium sulphate solution (prepared by dissolving 1.5 g in 100 ml distilled water with 4.6 ml of conc H_2SO_4) is added till the colour disappears. After one hour, the colour chances from red to yellow, depending on the concentration of iodine present. The control solutions are also run side by side under similar experimental conditions for comparison.

Nitrogen

The elements, nitrogen and phosphorous, are known as bio-stimulants. Nitrogen is an essential building block element in the synthesis of protein. Therefore, its data are of paramount importance in evaluating the nutritional status of the water bodies. Four forms of nitrogen are present in the sewage and consequently in the receiving aquatic ecosystem. Freshly polluted system, especially by sewage contamination, shows higher concentration of ammonia nitrogen which in an aerobic environment changes to nitrites and then to nitrates. Barth (1953) calculated that, for the conversion of ammonia into nitrate, 4.5 mg oxygen for one mg ammonia oxidised, would be required.

Nitrite nitrogen is relatively unimportant in water pollution studies, because it is unstable and easily oxidised into the nitrate form. Nitrate nitrogen is an indicator of past pollution in the process of stabilization and seldom exceed 0.1 ppm in surface waters. Nitrate nitrogen as NO_3 should not exceed 45 ppm in drinking water, because of its serious and occasionally fatal effects on infant.

High concentration of ammonia nitrogen indicates that the pollution is fresh. With ageing, the concentration of albuminoid nitrogen goes on increasing while free ammonia concentration decreases in more or less the same proportion. 0.01 ppm of free NH_3 and 0.1 ppm of albuminoid NH_3 concentrations are the maximum allowable limits for water used for public consumption. Ammonia is one of the major constituents contributing to the toxicity of municipal waters. At high pH free NH_3 and its hydrolysis products are toxic in low concentration while at low pH, they are converted into ammonium ions (NH_4^{+1}) which are comparatively non-toxic.

Nitrate nitrogen

The increasing application of fertilizers in agricultural lands results in water pollution due to the leaching of nutrients like N and P into the drains. The presence of these pollutants only in about 0.3 ppm of nitrate nitrogen concentration and 10 ppb soluble P in water, are needed to support the excessive production of algae bloons and aquatic weeds causing undesirable colour, taste and odour, there by degrading the aesthetic quality of surface water bodies. Further, it has been estimated that the nitrate-nitrogen content varies from 0.04 to 5.0 ppm in the rivers of the world.

A survey of the well waters in low lying areas around Patna, Bihar are elsewhere in India reveals that in certain wells, the nitrate-nitrogen was as high as 240 ppm which is considerably above the WHO limit of 10 ppm nitrate as nitrogen. Such high amount of nitrogen in well water is enough to produce disease known as methemogiobinemia in infants. 10 ppm of nitrate nitrogen concentrations interact under acidic condition with secondary amine compounds present in the water to produce nitrosoamines which are well known carcinogen. Another serious repercussions of high nitrate nitrogen concentration, is that non-pathogenic bacteria which live freely or in symbiotic association with plants, may get destroyed, resulting in a serious break down in the self-purification process of the aquatic ecosystem.

Procedure for determining free ammonia and albuminoid ammonia nitrogen estimation.

250 ml of water sample is usually distilled off in two different stages for measuring nitrogen in the above mentioned two forms. Accordingly, 250 ml of water sample is first distilled off at pH higher than 11.0. Such high pH value is achieved by adding 2 pellets of sodium hydroxide (NaOH). The distillation is continued, till 50 ml distillate is obtained (A). Thereafter 50 ml of alkaline potassium permanganate solution is added to the remaining constituents in the distilling flask and the distillation is continued till another 50 ml of distillate is collected in another nesselar tube (B). Both the distillates A and B, along with a series of control samples in separate nesselar tubes are arranged on white porcelain based stand. To each of the nesselar tubes, 2 ml nesselar reagent (K_2HgI_4) is added. Nesselar reagent is prepared by dissolving 35 g KI in 100 ml distilled water to which saturated solution of $HgCl_2$ is added .

Subsequently, 120 g of NaOH is added, dissolved and diluted to 1 litre. Thereafter, the colour developed is matched with standard solutions. Supposing, if the concentration of ammonia is found to be 0.01 mg in the examined water sample after comparing it with the standard solution of ammonia, the sample examined is 250 ml therefore 1000 ml of water sample will show 0.01 x 4.0 ppm.

4-hours oxygen consumption test

This test is used to measure the easily decomposable organic content present in water and waste samples. Mild oxidising agent like potassium permanganate ($KMnO_4$) in acidic medium at 37^0 C is used to oxides the organic materials. 125 ml of water sample is taken, to which 25 ml of N/ 80 potassium permanganate solution is added along with 10 ml of N/2 sulfuric acid. The whole content is kept at 37^0C for 4 hours. At the end of the incubation period, one spoon of solid potassium iodide (Kl) is added and titrated against N/40 sodium thiosulphate ($Na_2S_2O_3$) solution, using starch as indicator.

Bio-chemical oxygen demand (BOD)

BOD represents the intensity of bio-degradable matter remaining in the stream at any time and DO (dissolved oxygen) shows the ability of the stream to purify itself through bio-chemical process. BOD tests show the amount of molecular oxygen required by bacteria to reduce the carbonaceous materials. It is a bio-assay procedure that measures the oxygen consumed by living organism, while utilizing the organic matter. Bio-chemical oxidation is a slow process and therefore, 20 to 30 days are required for the complete degradation of the waste. Oxidation is achieved upto 95 to 99 percent in 5 days in case of glucose, about 65 percent in case of sewage, while 40 to 50 percent of the organic matters present in any other industrial waste water samples during the same incubation period. Both, the 20^0C incubation temperature, used as an average value for slow moving water bodies in temperature climate and easily reproducible in an incubator, and the 5 days incubation period, were chosen by British Royal Commission for sewage disposal because none of the rivers in England has a reaching time to the sea of more than 5 days.

Estimation

For the BOD test to be performed the sample of water from rivers are taken in duplicate without going for any dilution. In case of strong sewage or industrial wastes, the samples drawn are suitably diluted so that the DO present in the sample may not get completely exhausted during incubation period. Dilution is thoroughly aerated and enriched with sewage water nutrients for the microbes. 5 to 10 ml of per litre of diluted water is added as seeding material, prior to subjecting to the process of aeration. After suitably diluting the strong waste water sample in duplicate, one is put under incubator at 20^0 C for 5 days and the other set is put for DO dermination. After the incubation period is over, DO contact in the sample is again estimated.

Chemical oxygen demand (COD)

COD test shows the oxygen equivalent of the organic matter that can be oxidised by using a strong oxidising agents e.g. potassium dichromate in acidic solution, at elevated temperature, for two and half hour. Silver sulphate is used as catalyst to oxidise the oxidation resistant organic compounds. In case where BOD test fails, COD succeeds. For example, when, municipal or industrial waste containing organic compounds that are toxic to biological life, produces erroneous BOD results. COD value is in general always higher than that of BOD. The ratio of COD/ BOD shows the types of treatment required for the particular waste waste water sample.

(i) When BOD : COD -0.6, it means that the waste is rich in putreascible matter. Therefore, the best treatment for such type of waste is biological one.

(ii) When BOD : COD is between 0.3 to 0.6, it means that the waste water requires acclimatization, prior to subjecting it for the biological treatment.

(iii) If and when BOD : COD = 0.3, the chemical treatment is the best for such waste.

Estimation

COD test is done by adding 20 ml of 0.25 N solution of

potassium dicromate along with 75 ml of conc sulphuric acid to 100 ml of waste water sample, in a digestion flask of 250 ml capacity. About 5 g of silver sulphate as catalyst is added to the reaction mixture.

The whole set along with reaction mixture is refluxed for two hours. Thereafter, it is cooled under tap water and then titrated against 0.25 N solution of Mohr's salt (ferrous ammonium sulphate) using 2 to 3 drops of foreign (prepared by dissolving 1.4 of 1:10 phenonthroline monohydiate together with 0.69 g of ferrous sulphate in 100 ml distilled water) solution as indicator, till the end point is reached. A blank sample is also digested in similar experimental condition. The value of COD is computed in ppm as given below

$$\text{COD (ppm)} = \frac{(a-b)\ c \times 8000}{\text{volume of sample (ml)}}$$

where, a = vol. of titrant used (ml) in blank,

b = vol. of titrant used (ml) in sample

c = factor of the titrant

Phosphate as p

50 ml each of water sample along with a suitable control are taken in a series of of nesselar tubes. To each of the tube, 2 ml of ammonium molybadate solution prepared by dissolving 25.0 g ammonium molybadate in 175 ml of distilled water (soln A). In another vessel, 280 ml of conc. H_2SO_4 is added to 400 ml distilled water (Soln. B). Solutions A and B are then mixed together. Thereafter, five drops of freshly prepared stannous chloride solution of 2.5 percent strength are added. The whole content of the tube is mixed, which result in the development of blue colour in water as well as in control sample. The visual colour comparison riveals the concentration of phosphate as (p) in ppm by multiplying with 20.

Silicate as SiO_2

Similar method as that for phosphate estimation is carried out for silicate estimation in water sample by adding 1.1 HCI and 2.0 ml of ammonium molybadate solution. The whole content is thoroughly mixed and kept for five to ten minutes. Thereafter, 1.5 ml of 10 percent oxalic acid is added. The yellow colour thus developed, is compared with those of control solutions and accordingly calculated by multiplication with 20. The result is expressed as SIO_2 in ppm.

Nitrate as nitrogen

To 50 ml of water sample, 2 ml of 30% sodium chloride solution along with 10 ml conc. H_2SO_4 are added and mixed gently. Brucine reagent (0.5 ml) is then added. The whole reaction mixture is put on water bath for 20 minutes, the colour developed is compared visually with suitable control solution and calculated by multiplying with 20. This gaves the value for nitrate as nitrogen in ppm.

Sulphate as SO_4

100 ml of water sample with not more than 40 ppm of a sulphate is taken in nesselar tube. Similarly, a series of control, samples is taken in different nesselar tubes. To each of the tube, 5 ml of conditioning reagent is prepared by dissolving 76 g NaCl, 30 ml conc. HCL and 100 ml of 95 percent ethyl or isopropyl alcohol in 300 ml distilled water. After thoroughly mixing the whole content, one spoon (say 200 mg) of barium chloride ($BaCl_2$) crystal is added. The turbidity developed is visually matched with that of the control solution of the sulphate within 5 minutes of the $BaCl_2$ addition.

Bacteriological parameters

Most probable number (MPN)

MPN count of water samples, drawn from the respective

sources is determined by using multi-tubes-dilution techniques. In the procedure, the McConkey lactose bile salt media are filled up in test-tubes of different sizes and sterilized under 15 lbs pressure per square cm at 140°C for 15 minutes. Thereafter, the tubes are arranged in four to five series, with three tubes in each series, for each sample of water to be examined. To each of this set, water sample is transferred through sterilized pipettes.

Each series contains different volume of original water samples i.e. the three tubes of the first series are inoculated each with 10 ml, the second series of three tubes with 1.0 ml each, while 0.1 ml each for the third and so on up to the fifth series by adding 0.01 ml and 0.001 ml respectively to all the three tubes of fourth and fifth series. Thereafter, all the sets are put at 37 x 2°C for 48 + 2 hours. At the end for the incubation period, each tube is examined for gas production in the durhms tube. The tubes having gas production are marked as positive ones and those without gas production are discarded. Numerical values of MPN count present in 100 ml of water sample are enumerated with the help of the already worked out chart, based on the formula developed by Thomas.

Fecal coliform Count (FC)

All the positive tubes of MPN count are sub-cultured in duplicate, in brilliant green lactose broth (BGLB) media, One set is put at 37° C for 48 hours and other set at 44 x 0.2°C for 24 hours. The values derived from the former set how the confirmatory number of MPN count/100 ml where as the latter set shows the numerical values for the fecal coliform counts present in 100 ml of water sample under investigation.

Escherichia coliform count (E.Coli)

All positive tubes of either fecal coliform or MPN are

further sub-cultured in a series of three test tubes of Mc Conkey bile-salt lactose broth and put for incubation at 37⁰C for 24 hours. Again, the gas production in each tubes is examined. The positive tubes are allowed to proceed for further treatment at follow:

The turbid inoculum form each of the +ve tube is inoculated through loop of standard size in the McConkey agar plate and put for incubation at 37⁰C for another 24 hours. At the end of the specified time, the well developed pinkish, flat and transparent colonies are further inoculated in peptone water and McConkey media.

The whole set is again put at 37⁰C for another 24 hours. The former, on producing positive reactions with Kovac's reagent and the latter, showing gas production, are utilized for the numerical assessment of escherichia coliform count, present in 100 ml of sample under investigation.

Kovac's reagent is prepared by dissolving 5 g paradimethly amino benzaldehyde in 75 ml of either isoamyl alcohol or absolute alcohol and 150 ml of conc. $H_2 So_4$

Fecal streptococci count (FS)

Fecal streptococci count is measured by adopting similar methods as for MPN and other counts, as discussed earlier and expressed per 100 ml of water samples. The only difference is the use of nonlactose media in the case of F. streptocci estimation. Therefore, azide dexirose broth (ADB) media are used in the tubes arranged in 4 series of three tubes each. To each of the tube, water sample in different dilution i.e. 1.0, 0.1 and 0.01 ml are inoculated and put at 37⁰C for 48 hours of period. The tubes showing well marked turbidity are considered as positive tubes.

Further, the positive tubes are subcultured in ethyl violet azide (EVA) media and put to incubation for another, 48 hours at 37 x 2^0 C. The formation of ring at the bottom of the tube, confirms of fecal streptococci in the sample. The numerical values are ascertained again by using the Thomas chart, as used in evaluation of the MPN and E. coliform counts.

4 Systems Approach to Environmental Problems

Man's environment has suffered severe degradation because of failure to evaluate problem areas on a total systems basis and because of unforeseen consequences. Rachel Carson called attention to instances of unforeseen results from the use of chemical pesticides. Biological concentration of DDT illustrates the interactions in the biological systems. The wide use of nondegradable detergents dramatically demonstrated channels of population and also resulted in objectionable foaming.

The present handling of solid wastes in urban areas reveals many deficiencies in failing to consider the handling in the dwelling, collection and transportation and the disposal, recycling, or reclamation as a complete system. The disposal method, incineration, hog feeding, or land filling, may dictate the segregation and handling at the dwelling. Economic, legal and political factors, and public attitude may be decisive in selecting a method of disposal.

The collection of all sewage at a single plant may make possible the construction of a more efficient treatment plant, but the increased absolute waste load, even after treatment at the high efficiency, may be more than the stream can assimilate. Discharge of the wastes at several points, even at lesser efficiencies in removal, might have been possible with no difficulties. Failure of highway planners to consider noise, air pollution, and social disruption from highway alignment has resulted in citizen action to block proposed "improvements."

How can we avoid such pitfalls in environmental planning! The basic land use pattern which is adopted, or allowed to develop haphazardly, is a significant determinant for the future needs and problems of the area. The livability and character of development is related to the density and distribution of the population. Once such things as to type of dwelling (individual detached homes or high rise apartments), transportation system, allocation of land uses, etc. have been allowed to develop, major changes are difficult and expensive to accomplish.

Environmental planning should be considered at the time of land use planning and development. The Department of Agriculture can supply data on soils which will reveal areas most suitable for parks, for residential lawns, or for absorbing the liquid effluent from septic tanks. Geological surveys will show the suitability for foundations, underground water resource potential, and areas suited for landfall waste disposal. New subdivisions in urban areas seldom consider these factors and do not make adequate provision for solid waste and sewage disposal. In densely populated areas,sewage collection and disposal should be provided initially. Septic tanks sometimes fail after a few years of use by city dwellers accustomed to sewers. Solid waste handling in high rise apartment and office buildings requires as much care in engineering planning as climate control, structure, and other factors.

Functional use of spaces inside and outside the dwelling unit for satisfaction of human needs, both individual and family, must be considered as an interrelated system. Spaces for individual privacy, family activities, and community social interaction are required. Stimulating variety is essential to human development and preservation of individuality.

Frequently, in the past, aesthetics and environmental problems such as noise and vibration, traffic, air pollution, and waste disposal, have been neglected by industry. As a consequence, past practice has been to separate industrial areas by buffer zones. The result has been tedium and stress of long travel time to work and increased pollution from automobiles. The imaginative approach to a pleasing architectural design of modern industrial buildings provisions of landscaped open space, are the solution of commercial and industrial development in the residential area. Rigid performance standards are essential in such circumstances.

Waste disposal facilities have been similarly shunned as neighbours. Landfills, incinerators,and sewage treatment facilities have given problems where operation was faulty. With attractive architectural treatment and proper operation, adequate facilities become more acceptable.

Analysis of an environmental problem on a systems basis will result in a better solution of that problem. However, a piecemeal attack on one problem after another will not necessarily result in a total environment which maximizes benefits to man's health. There is interaction between areas. A frequent illustration is that of an industrial hygiene problem. It may be solved by installing an exhaust system, only to create an air pollution problem. The air pollution problem is solved by installing a liquid scrubber, only to create a water pollution problem. The water pollution problem is solved by precipitating the pollutant. Now, what does one do with the solid waste?

The land, water and air resources on earth are limited. As population increases,pollution problems are intensified. As the magnitude of pollution potential increases, higher and higher efficiencies in treatment are required to maintain the status quo. Soon 100% recovery of pollutants may become necessary. Under such circumstances, the prevention or elimination of waste becomes critical. Measures to reduce the amount of the waste are next in importance. The irreducible wastes remaining would present no disposal problem if they could be recovered and recycled or put to a useful purpose as raw material or to reclaim land as in the case of strip mines. The interrelatedness of the various areas of the environment is again shown by the necessity to consider the water pollution potential of land reclamation projects using waste products.

It has been shown that the evaluation and solution of environmental problems must consider biological interactions,human stress,land use patterns, and inter relatedness of environmental areas. In addition, political and economic influences and consequences may be critical and decisive factors. Many environmental problems, e.g., water resource development, land use, and waste (air, water, solid) management require "problem-shed," or regional, approaches to provide effective control. Political innovations may be required to effect regional environmental management.

The importance of mathematics in all branches of science, technology, engineering, socio-economics, etc. is known to everyone of you. Differential equation and integration are so common in all applied sciences that at every stage we have to turn to mathematics for solutions. The reason is that mathematics is an exact science. In contrast biology, for instance, is still regarded as an inexact science. This is because biology deals with life science and it started as purely descriptive science. It is inherently complex and enough discoveries have not been made to bring precision of the type mathematics has. Today in inexactness in biology is due to our lack of knowledge.

Environment, it is defined as "the sum of all external conditions and influences affecting the life and development of organism. "This area is thus a conglomerate of all basic and applied sciences, engineering, socio-economics and law. Added to it is the fact that environment is for the living system, which naturally brings in a measure of inexactness in the areas of environment.

The environment of region, country, state, district, city, town and village is influenced by several factors, because there is a high degree of location-specificity in environment, on account of an all round diversity starting, with ethnic considerations, and going over to geological, geographical, climatic, floristic, faunistic, social, economic and cultural considerations.

India is a country where at this stage, centuries are co-existing. We have both near stone-age and near spaceage, cultures running parallel. We have human beings existing as part of nature in perfect harmony with it; taking from nature only what they require and nothing more and existing as a part of the eco-system, as one of the species. At the other end is the socio-economic system of the human kind where he is a co-creator and in conflict with nature and is dominated by considerations of greed. There are all manner of integrating situations in between. We have thus problems that afflict both developing and developed countries. This is true of all phases of national developments. The situation in India is relevant to both developing as also to the developed world. Today, our thinking about environment comes from the West. Historically, our understanding of environment was holistic.

The whole universe together with its creatures belong to the Lord (Nature). One can enjoy bounties or nature by giving up all greed. Implicit in the statement is that no creature is superior to others and human being should not have absolute power over nature, let no one species encroach on the rights and privileges of any other species.

This element of sustainability is ingrained in this because the emphasis is to use nature without greed. Once an element of greed enters, exploitation starts and we cease to utilize nature for the good, the benefit and the welfare of our fellow human beings. Sustainable development is therefore, a concept of good and sound economic growth, a growth that can be maintained indefinitely with no or minimal damage to the environment. Good environment will ultimately beget good economics. The next question arises as to what are the factors that can promote sustainable development. These are :

(a) Population stabilization and health care;

(b) Integrated land use planning and watershed management. Ensuring adequate use of cropland, woodland or grassland for food, fuel, timber and fodder;

(c) Revegetating marginal land and greening the uncultivated half;

(d) Water pollution control in rivers;

(e) Air pollution control in industrial pockets;

(f) Non-polluting renewable energy;

(g) Waste recycling and reuse;

(h) Conservation of biological diversity;

(i) Human settlements without congestion and slum improvement; and

(j) Environmental education and awareness.

The elements of the life support system, namely, air, water, land, flora and fauna and the incoming sunlight form an interacting system. Sustainability of development depends on such a complex system. Even problems like the carrying capacity of our land has number of dimensions like relationship between land area, topography, population growth, and production/ productivity of food, fodder, fuel, timber, fertilizer, fibre, medicines, etc. For answering such questions, it is necessary to develop a systems approach. Furthermore, Mathematical Modelling is one of the important methods to resolve such complex situations. The other approaches could be PERT (Programme Evaluation Review Techniques). CPM (Critical Path Method), OR (Operation Research) and so on.

Systems Approaches have acquired a significant place in modern technological development because these help to look at the problems in totality particularly for situations where different components give an impression of not working in cohesion, thus enabling to deal with whole system involving many parameters; operational mechanisms together with available resources. All these aspects need to be taken up in an integrated approach. Systems Approach has a wide scope as it helps to resolve problems in optimal utilization of resources at minimal cost.

For instance, water quality models can be prepared about surface water, estuaries, lakes and large rivers in selected stretches with use of water quality parameters. Equally important are models on ground water in which impact on ground water quality by solid waste disposal which leads to leaching of chemicals through rain water can be worked out with reference to rainfall pattern, soil porosity and height of ground water table. This would also include monitoring of parameters like chloride, (Phosphate, COD), BOD.

Similarly, air quality modelling can be undertaken with reference to pollutant from growing industrialization. We need impact assessment of environment and therefore, short and long range transport of pollutants from industries in different terrains, meteorological conditions and in different industrial complexes. Validation of models would be needed by actual monitoring for operational and regulatory purposes as also air quality standards based on effects of pollutants on human beings, animals, plants and historical monuments.

Modelling should also be attempted regarding CO_2 from fossil fuels and biomass burning together with effect of its increasing concentration on climate, urbanization and acid rain.

A Case Study: An example where systems analysis and mathematical simulation has been useful in European part of Russia. Here, most of the rivers flow north and account for 84 per cent of country's water resources. However, Caspian Sea Basin in South accounts for only 16 per cent of water resources but 80 per cent of the production system is located in this area. There was a suggestion for partial diversion of north productivity of the aquatic and terrestrial biological system. However, before undertaking the projects,the negative consequences, of such a diversion as also estimate the maximum permissible volume of diversion had to be worked out. In this exercise,over 100 institutes were involved from 1976 to 1982. The study did not favour the project.

Application of System Analysis Techniques to environment is not new to the country and several applications already exist. These are concerned with the well known traditional areas. Nearly 50 institutions/ organizations are today involved in different types of application of Systems Analysis of environmental problems. Some of the areas where systems analysis has been applied in the country are :

(a) Water Supply and Water Quality Planning for, (i) Urban water supply system, (ii) Rural water supply system, (iii) Water quality planning and management with reference to

(b) Estuarine water quality models BOD, DO coupled system models.

(c) *Waste Water Transport and Disposal System for,* (i) Drainage system in multistoreyed buildings, (ii) Waste water transport, (iii) Waste water treatment systems.

(d) *Solid Waste Management for,* (i) Solid waste collection system.

(e) *Air Quality Planning and Management for,* (i) To determine optimum stack height emission rate, (ii) To compute long term and short term variation of the distribution of air pollutants and particulate deposition arising from elevated point sources; (iii) To determine the average pollutant concentration over a specified period of time; (iv)To allocate the maximum pollution emission rates from various sources in a region.

(f) *Land and Water Resources Development for,* (i) Degradation of soil fertility due to increase in sodium concentration; (ii) Physical loss through accelerated erosion; (iii) Impact of conversion of good farm land into reservoirs.

(g) *Forest Eco-systems for,* (i) Tropical deforestation, (ii) Erosion control, (iii) Waste water treatment and disposals of land fills, (iv) Urban and rural settlements, (v) Ecological modelling.

A major requirement in modelling is to have good

quality data which is accurate and reliable for processing before use in models.

For this purpose we have established Environmental Information System (ENVIS) using LSI 210 Computer System which is hooked to the Cyber 170/720 systems of NIC. The ENVIS is geared to the needs of five types of target groups like, policy planners, decision makers, research workers, administrators and the public. Information of environment is constantly in demand. ENVIS is a decentralized system using the distributed network of data bases and ensuring integration of national efforts in environmental information collection, storage, retrieval and wide dissemination by avoiding duplication of efforts.

The Department of Environment is also the National Focal Point (NFP) of INFOTERRA (An International Referral System for Sources of Information) network of United Nations Environment Programme (UNEP). INFOTERRA provides a referral sources on environment from a global computerized file of about 100,000 sources in response to specific queries from both national and international users. NEP has registered more than 530 national sources in the global network. A thorough updating of these sources in the country has been completed recently.

ENVIS centres in different subject areas have been set up. Work in the ENVIS centres relating to their respective subject areas of environment are referred for supplying the information. Some specific activities taken up in these ENVIS Centres include :

(a) Water and air pollution

(b) Toxic chemicals

(c) Coastal and offshore ecology, environmental mapping in Eastern Ghats

(d) Eco-toxicology; biodegradation; environmental impact assessment

(e) Environmentally sound and appropriate technologies

(f) Renewable energy and environment

(g) Non-government organization, Parliamentary questions and media as related to environment

(h) Western Ghats; biological diversity

(i) Environmental management in Madhya Pradesh.

ENVIS will also help the proposed National Environmental Monitoring Organization (NEMO) to be established soon. The objectives of NEMO are :

(a) To build up base line information on the state of Environment in India by monitoring the abiotic and biotic components of the eco-systems with a view to :

   (i) Introducing environmental dimensions in economic and technical planning of development projects ;

   (ii) building up capability to assess the environmental impact of proposed development projects and to propose environmental safeguards;

   (iii) creating a general awareness of the need to preserve the environment using formal and informal educational programme for different levels of the communities.

(b) Acquirement of capabilities listed above would lead to the goal of providing strong data base with sufficiently accurate information content to evolve environmentally compatible development programme with concomitant preservation of environmental quality and the productivity of natural systems. To attain these capabilities, it is proposed to set up an institutional mechanism in the form of a National Environmental Monitoring Organization which will be entrusted with tasks listed below :

(i) To prepare an inventory of the uses of surface waters, within the territorial limits of India, stretch by stretch and identify the highest quality use and prepare the water use maps based on the highest quality use.

(ii) To monitor the quality of the surface waters within the territorial limits of India and to develop water quality maps.

(iii) To monitor the ambient air in areas free from pollution and in areas under the influence of varied degree of pollution.

(iv) To monitor the quality of land in specific areas.

A few suggested examples are :

(a) Intensely cultivated
(b) high pesticide used
(c) high fertilizer used
(d) endemic
(e) material (hazardous dumping)

(c) To conduct post-mortem investigation and assess environmental impact of projects measuring changes in abiotic and biotic components of the eco-systems in the neighbourhood of the project.

(d) To monitor the abiotic and biotic components of the eco-system in specific neighbourhood of developmental projects in pre-implementation, during implementation and post implementation stages.

(e) To promote research and development for need based environmental monitoring.

(f) To undertake nation-wide monitoring in areas of special interests such as National Monuments, National Parks, Religious or Cultural Sites, Tourist Resorts, etc.

(h) To co-operate with other agencies in specific studies to monitor the earths resources or natural calamities and other organizations having specialist capabilities for collection and analysis of data.

(i) To co-ordinate and undertake special mass education programmes, induct mass participation and arouse a general awareness among the people for the need to preserve the environment.

The aforesaid tasks if undertaken successfully by the National Environmental Monitoring Orgnization would serve the following important purposes :

(a) Provide base line reference data and requisite data and feed back for the assessment of environmental impacts for different development projects

(b) Provide as a service agency the necessary inputs to experts or the government or other users for evaluation of the current state of the environment and help the users in adopting appropriate environment management policies in some suggested areas such as :

    (i) industrial siting

    (ii) infrastructural development

    (iii) water resources development

    (iv) mineral development including fossil fuel

    (v) coastal management

(c) Provide data on assimilation capacity of sinks and thereby help user agencies in formulating measures and in assessing the effectiveness of such measures through the indication of environmental quality improvement.

(d) Serving as a data bank on environmental monitoring for all users and evolution of regional and sub-regional data-based environmental management systems.

(e) Serving as an Organisation to educate the general public against the hazards of pollution and making them participate willfully in pollution control in spite of reverse economic implications.

(f) Providing early warning services about environment hazards.

There would be nearly 150 Field Centres located throughout the country and would be aggregated under 32 sub-zonal centres distributed under 5 zonal centres at Delhi 3, Lucknow 6, Bombay 5, Madras 8 and Calcutta 10. The Headquarter would be involved in receiving collating, processing, interpreting and retrieval system.

Data management involving data abstraction, formulating, validation can be efficiently handled with the help of computers. Most institutions today have in house computers with facility to use outside computers also.

Water is absolutely essential not only for survival of human beings, but also for animals, plants and all other living beings. Further, it is necessary that the water required for their needs must be good and it should not contain unwanted impurities or harmful chemical compounds or bacterias in it. Therefore, in order to ensure the availability of sufficient quantity of good quality water, it becomes almost imperative in a modern society, to plan and build suitable water supply schemes which may provide potable water to the various sections of community in accordance with their demands and requirements.

The provision of such a scheme shall ensure a constant and a reliable water supply to that section of the people for which it has been designated. Such a scheme shall not only help in supplying safe wholesome water to the people for drinking, cooking, bathing, washing, etc. so as to keep the disease away and thereby promoting better health; but would also help in supplying water for fountains, gardens, etc. and thus helping in maintaining better sanitation and beautification of surroundings, thereby reducing environmental pollution. Besides promoting overall hygiene and public health, it shall ensure a safety against fire by supplying sufficient quantity of water to extinguish it. The existence of such a water supply scheme shall further help in attracting industries (since industries require large amounts of water) and thereby helping in industrialisation and modernisation of the society, consequently reducing unemployment and ensuring better living standards. Such schemes shall, therefore, help in promoting wealth and welfare of the entire humanity as a whole.

Demand of water

India is likely to have a population of around one thousand million by the end of the century, out of which the urban population will be around for hundred million. This would entail a large responsibility on governments and civic bodies to provide all communities with safe drinking water as also to plan for industrial and other needs. Water is required everywhere, without which neither the life nor any development is possible. Public water supply is among the most important needs which mainly caters to the following uses:

(i) Domestic: The use of water for drinking, bathing, watering of lawn and gardens, air conditioning, flushing of toilets, livestocks etc, is designated under domestic use.

(ii) Institutional: Use of water in hospitals, educational institutions, hotel and restaurants hostels, cinema halls, offices, factories, railway stations etc. is considered as the institutional need of water.

(iii) Public purposes: Water utilized in street sprinkling, watering of public parks, display fountains and cascades, swimming pools and flushing of sewers is considered in the category.

(iv) Fire fighting: Water used in fighting fire menance in both the rural and urban areas is usually considered as the accidental need of the society.

(v) Industrial and Commercial: Apart from above, a substantial quantity of water is lost due to waste through public hydrants and among users.

Quantity of water needed

There are fixed Indian standards for the quantum of water to be supplied separately for the aforesaid basic needs, which are as follow:

Domestic and non-domestic need of water

The following parameters have been recommended depending on the habits and standard of living of the people, provision of flushing systems in residential building and above all the feasibility of supply.

(i)	For communities with population upto 10,000	70 to 100 litres/capita
(ii)	For Communities with population 10,000-50,000	100 to 125 litres/capita
(iii)	For communities with population above 50,000	125 to 200 litres/capita

Institutional need of water

Requirement of water earmarked for different institutions as enumerated below, will preferably be in addition to the requirements described earlier.

Fire fighting need of water

Requirement of water for fire fighting is recommended generally for communities with a population above 50,000. The provision, for this purpose, in kilolitres/day is computed from the formula 100/p where, p is the population in thousand. High rise buildings should have separate fire storage.

Table 1

Quantity of Water Needed by the Institutions in India

	Institutions	*litres/capita/day*
(i)	Hospitals (including laundry)	
	(a) No. of beds below 100	340
	(b) No. of beds above 100	455
(ii)	Hotels	1180
(iii)	Hostels	180
(iv)	Nurses homes and Medical quarters	135
(v)	Boarding Schools/Colleges	135
(vi)	Restaurants	70
(vii)	Air and sea ports	70
(viii)	Junction Stations and primary intermediate Stations	45
(ix)	Terminal Stations	45
(x)	Day Schools/Colleges	45
(xi)	Offices	45
(xii)	Factories	45
(xiii)	Cinemas/concert halls and Theatres	15

Industrial need of water

The water needs for different industries vary widely. Proper assessment will therefore have to be made keeping in view the possible future expansions. As for small industries, other than factories, located inside a town, the per capita supply already enumerated earlier is expected to be sufficient.

Table. 2
Quantity of Water Needed by Various Industries

Industry	*Unit of production*	*Kilolitres per unit*
Automobile	Vehicle	40
Distillery	Alcohol (in kilolitre)	122.170
Fertilizer	Ton	80-200
Leather	100kg	4
Paper	Ton	200-400
Paper (Special quality)	Ton	400-1000
Straw Board	Ton	75-100
Petroleum Refinery	Ton	15-2
Steel	Ton	200-250
Sugar	Ton (can Crushed)	1-2
Textile	100 kg (goods)	8-14

Sources of water

Water is widely distributed in nature and occurs in number of forms e.g., solid, liquid and vapour. There are two major sources of water viz., surface and ground. Brief account for each of them is given below:

Surface water

That part of rain water, which flows over the ground to feed rivers, brookes, streams, lakes, ponds etc. is known as run-off or flow-off water.

Rivers and streams are the important surface water resources, through which public water supplies are managed. Quality of surface water depends upon the season of the year, nature of soil and the habitation in the catchment area. But in general, surface water is impure and becomes contaminated where the river flow through thickly populated and industrialised areas. Lakes, natural and artificial, also show similar qualities as in rivers and streams but they, being quiescent, show less pollution both physically and bacteriologically.

Ground water

Gravity and artesian springs, deep and shallow wells, and infiltration galleries under sandy river beds with a good depth of sand, are the common sources of groundwater. In geological term, the same has been defined as the water that reaches the zone of saturation. It is generally pure when drawn from greater depths as in the case of deep tube wells except that the water may be hard. Public water supplies from groundwater sources are economical as well as less complicated.

Since, the groundwater is found at variable depths hence, exhibits variable properties. Some of the important terminologies about various types of groundwaters are explained below. Water, which has never been a part of the hydrosphere, is known as Juvenile water. Similarly, a water that has recently been involved in atmospheric circulation is known as meteoric water. The thermal spring water is the transformed form of juvenile into the meteoric water. Besides, there are waters which have been out of constant with atmosphere for millions of years and are known as connate water. The water that has been in association with rocks during their metamorphism is termed as metamorphic water.

The water that is derived from magma is known as magmatic water. The magmatic water found at shallow depth is known as volcanic water and that at considerable depth as plutonic water. Many terms are frequently used in connection with the development of groundwater resource management and therefore, their explanations are presented briefly here as follows:

The locus of levels to which water rises in free communication with soil voids is termed as water table. The water table remains constant so long as the withdraw and recharge of the system is maintained in equilibrium. The water bearing strata, capable of yielding sufficiently large quantity of water, so as to be comfortably utilized by the community is termed as aquifer. An aquifer held under pressure by an overlying impervious stratum is known as confined aquifer and as such the water is withdrawn with the help of pump. The aquifer subjected to hydrostatic pressure (as actually happens in case of spring), creates a condition, widely known as artesian, where the flow of water is spontaneously maintained without any pumping device. Water used for the drinking purposes is known as potable water, which must be sparkingly clear, odourless, neither very hard nor too soft and be free from bacteria. Saltish water often fails in quenching the thrust, hence must satisfy all the physico-chemical and bacteriological quantity criteria for human consumption. Waters, on the other hand, used to feed boilers, for washing, and many other non-domestic and industrial purposes are termed as non-potable water.

The fundamental difference between potable and non-potable water lies in the degree of treatment given to them for purification. The waters which are ideally balanced in chemical composition and free from microbial contamination are termed as wholesome water. The suitable treatments needed to make the water wholesome have been thoroughly discussed in subsequent page.

General Impurities Present in Water

Both inorganic and organic impurities are present in the soluble, insoluble and colloidal forms in the water with their corresponding effects, are shown below:

(i) Suspended impurities

(a) Bacteria and Virsus: Some of them cause disease.

(b) Algae, protoza and fungi: cause odour, turbidity, colour and clogging of pipes.

(ii) Dissolved impurities

(a) Salts

Calcium: Bicarbonate and carbonate-Alkalinity and Hardness

Magnesium : Sulphate-Hardness

Chloride : Hardness, Corrosive to boilers

Sodium : Bicarbonate and Carbonate-Alkalinity

Sulphate-Forming in boilers

Chloride Brakish taste

Iron : Oxide-Bad taste, red water, discolouration of clothes corrosive of metals, hardness

Manganese : Black or brown water, discolours of clothes, helps organic growth

Lead : Cumulative poison, causing lead poisoning

(iii) Gases

(i) Oxygen: Corrosive to metals.

(ii) Carbon dioxide: Acidity, corrosion

(iii) Hydrogen sulphide: Odour like rotton egg, acidity.

(iv) Nitrogen: Albuminion ammonia-Vegetable pollution, Free ammonia sewage pollution.

Table 3

Physico-chemical Limits for Drinking Water

Parameter	*Acceptance*	*Cause for rejection*
Turbidity (J.T.U. scale)	2.5	10
Taste and odour	Objectionable	Objectionable
Total dissolved solids (mg/l)	500	1500
Chloride (as Cl) mg/l	200	1000
Flouride (as F) mg/l	1.0	1.5
Calcium (as Ca) mg/l	75	200
Iron (as Fe) mg/l	0.1	0.1
Copper (as Cu) mg/l	0.05	1.5
Phenolic compounds (as phenol) mg/l	0.001	0.002
Toxic Materials		
Arsenic (as As) mg/l	0.05	0.05
Chromium (as hexavalent Cr) mg/l	0.05	0.05
Lead (as Pb) mg/l	0.1	0.1
Mercury (total as Hg) mg/l	0.001	0.001
Gross Alpha activity (pcl)*	3	3
Colour (platinum cobalt scale)	5.0	25
pH	7.0 to 8.5	6.5 to 9.2
Total hardness (mg/l) as $CaCO_3$	200	600
Sulphate (as SO_4) mg/l	200	400

Nitrate (as NO_3)mg/l	45	45
Magnesium (as Mg) mg/l	30	150
Manganese (as Mg) mg/l	0.05	0.5
Zinc (as Zn) mg/l	5.0	15.0
Anionic detergents		
(as MBAS) mg/l		
Toxic Material	0.2	1.0
Cadmium (as Cd) mg/l	0.01	0.01
Cyanides		
(as CN) mg/l	0.05	0.05
Selenium (as Se) mg/l	0.01	0.01
Polynuclear aromatic		
hydrocarbons		
Radio activity 0.2/µg/l		0.2/µg/l
Gross Beta activity (pcl)*		30

pcl* = pico curie, equal to 10^{-12} cl

Whereas 1 cl = 3.7×10^{19} atoms disintegration/sec (dps)

= 2.2×10^{12}dpm

Conventional treatment methods for public water supply

Conventional methods for water supply scheme mainly achieve the following purposes by involving a number of step. Some of them are given below:

(a) Removal of colour, odour and taste

(b) Removal of excessive hardness

(c) Removal of iron and manganese

(d) Maintaining flouride level

(c) Demineralisation if needed

(f) Chlorination.

The steps taken in raw water treatment is presented below:

Raw water → PC → A → Fl → POC

Where PC: Pre-chlorination

A: Aeration

F: Flocculation

Fl: Filtration

POC: Post-chlorination.

The sources and the quality of raw water determine the type of treatments to be undertaken. For example, if the water is drawn from deep tubewells, coagulation and filtration are not necessary. Only softening may be necessary where the water is hard, or aeration where the water contains excessive iron taste or odour. For surface water, flocculation and disinfection are essential whereas prechlorination and aeration may be necessary in specific cases, and especially during rainy season.

Aeration

The principle behind this method is to mix air with water to increase the oxygen content and remove carbon dioxide. Aeration is done to remove iron from groundwater and taste and odour from surface water. It is mostly done in the initial stage of water treatment. It is a simple process and is accomplished by the following methods:

(i) Cascades-Water is made to fall in a thin sheet down a series of steps like a fountain.

(ii) Spray Nozzles-Water is sprayed under pressure to open air through nozzles.

(iii) Waterfall or Multiples Tray Aerators-Water is discharged through a riser pipe and allowed to fall on a series of trays from which it falls either through small openings or over the edges.

(iv) Mechanical Aerators-These are mostly applied in waste water treatment, where a large amount of organic matters are to be oxidised, within a short period of time. This method is suitable also for the area where land is a limiting factor.

Coagulation and flocculation

Finely divided suspended matters, both organic and inorganic which cause colour and turbidity, do not settle easily through sedimentation in a short period, the object of coagulation is to convert these into easily settleable size, by reducing the zeta potential forces. The latter are repulsive forces which on reduction, bind together the colloidal particles. Thus, the aggregated molecules of suspended colloidal particles become large enough in size and settle down easily. When the coagulants are added to water, large size flocks are formed on rapid agitation. These flocks absorb colour and during their downward journey entrap suspended matter and bacteria. This process is known as flocculation.

The most common coagulant is aluminium sulphate or $Al_2(SO_4)_3$, $18H_2O$ and the next is ferrous sulphate, $FeSO_4$, $7H_2O$. Alum is preferred because it requires lesser skill and control. The reactions that take place are as such:

$$Al_2(SO_4)_3 + 18H_2O + 3\,Ca(HCO_3)_2 = 2Al(OH)_3 + 3CaSO_4 + 18H_2O + 6CO_2$$

In above reaction $Ca(HCO_3)_2$ represents the natural alkalinity of waters. Some waters which do not have sufficient alkalinity need the addition of lime. CaO (lime) react with water to form calcium hydroxide, $Ca(OH)_2$ which further reacts with alum as follows:

$$Al_2(SO_4)_3 + 18H_2O + 3Ca(OH)_2 = 3Al(OH)_3 + 3CaSO_4 + 18H_2O$$

In both cases the flock aluminium hydroxide is formed as

coagulation depends on the pH of water. Different coagulants may be suitable for different pH range. Alum is most commonly used in public water supply scheme as coagulant. The use of alum in the form of solid in water treatment process is not desirable at all, because as such, the natural buffering capacity of aluminised water becomes, locally upset, which in turn, produces excessively high turbidity.

Table 4

Commonly use Coagulants to Raw Water Treatment

Common Coagulants	*pH range*
Alum	6.5 to 8.0
Ferrous Sulphate	8.5 to 11.0
Ferric Sulphate	5.0 to 11.0
Ferric Chloride	5.0 to 11.0
Sodium Aluminate	6.0 to 8.5

Thus, for efficient coagulation and flock formation, optimum pH, adequate dose, rapid agitation and gentle transfer of flocculated water to selling tank is necessary.

Quality criteria for alum and lime as coagulants

Good quality of alum must contain alumina (Al_2O_3) in the range of 16 to 17%. Besides there must not be more than 0.7% of insoluble matter and 0.3% iron. The pH value of 10% aqueous solution of alum should vary in the range of 2.7 to 3.2 (IS 299-1975) otherwise natural alkalinity of the water to be coagulated would be severely damaged resulting in very high consumption of lime.

Relaxation has however, been granted by Indian Standard Institution. Government of India, by reducing the limit of alumina to 15% for the use as coagulant especially in public water supply purification process (IS 299-1980). In case of coal washery wastes the use of alum as coagulant does not achieve the desirable degree of coagulation. For effective coagulation of the suspended coal particulates, recently developed poly-electrolyte in very little dose of 0.3 to 0.7 ppm in combination with alum and lime is of great success. The poly-electroyte as such works as coagulant aid.

Lime needs to be of equally good quality otherwise effective coagulation is not possible. Good quality of lime must have the minimum 90% content of available lime as CaO (IS-1540-1978). Too high content of lime as CaO in lime sample is to good to be used in river water treatment rather is wasteful in the case where natural alkalinity of raw water is more than 80 ppm. Experiments have conclusively proved that lime within 78.6% as CaO is good enough to be used in water purification, especially where natural alkalinity of water ranges from 65 to 110 ppm. Further the content of CaO in lime is shown to be considerably reduced on ageing, especially when the same is stored in the open air, owing to the conversion of CaO into $CaCO_3$.

Approximately, 50% of its total CaO content is lost during one and half month storage in the haphazard manner. Insoluble matter present in the lime as silica is though permitted upto 1.0 per cent (IS 1540-11-1971), but practically of no harm, even if it exceeds a number of times. This is rather helpful in reason where the hardness of water is found usually in the range of 100 to 125 ppm. At the same time, it should always be in mind that the insoluble matters, other than silica, are not desirable, because as such, the treated water becomes even more turbid than the raw water itself. Therefore special care is needed in the identification of the types of insoluble matters present as the impurities in both alum as well as in the lime.

For quantitative analysis of alum and lime, well

equipped laboratory is needed. Special care in weighing the samples and making volumetric solutions is of vital importance in achieving the accuracy in the result. The details of the procedures for the estimation of alum and lime are described in the IS 299-1975 and IS-1514-1059 respectively.

Filtration

Filtration aids in the removal of taste, odour, colour, turbidity, iron and manganese and to a large extent of bacterial load. Three types of filters are usually used in water treatment plants. There are:

(i) Rapid stand filter,

(ii) Slow and filter and

(iii) Pressure filter. Relative merits and demerits for each of them are discussed as follows:

Rapid sand filter

Essential characteristic of rapid sand filters are high rate of filtration and consequently lesser area requirement and washing the filters through reverse flow of filter water through the filter beds. A rapid sand filter consist of a bed, supported by gravels of different sizes in layers. Filter and sand for a rapid sand filter should satisfy the following norms:

(a) Sand shall be hard, resistant and free from dirt of every description

(b) Effective size should range between 0.45 to 0.70 mm

(c) Uniformity coefficient must not be more than 1.7 and less than 1.3.

(d) Ignition loss not be exceed 0.7 per cent by weight

(e) Not more than 5 per cent of the sand be soluble in HCl.

(f) Silica content should not be less than 99.8 per cent

(g) Specific gravity to be in the range of 2.55 to 2.65.

The depth of filter sand layer should be between 60 to 75 cm. The standing depth of water above filter sand must be between 1 to 2 meters. The free board above water level should be at least 50 cm. Besides these, the operation and maintenance of rapid sand filter need the strict implementation of the following knowhow:

> The standard rate of filtration through a rapid sand filter is usually 80 to 100 lpm/m^2 (litres per minute/ square meter). High rate can be achieved by proper conditioning of water before filtration. For plants of 1000 md (million litres a day) capacity or more, 100 m^2 area is needed to be provided in two units of 50 m^2 each. Besides the under-drain consists of a grid of manifolds, laterals, headers covering the entire filter bottom to collect filtered water and also to distribute wash water. These are perforated and pipes of cost iron, asbestos cement, concrete or other material.

For perforated pipe under-drain system, minimum gravel size should be 2 mm and maximum size 25 mm. Depth of gravel should be 50 cm. Filter gravel shall be spherical, hard, clean and uniform in size. Gravel is placed between the sand and the under-drainage system, the lower size being at the top and higher at the bottom above the under drains. It is essential to wash the filter beds at regular intervals to prevent clogging. Washing is done by wash water under pressure, sent through the under-drainage system into the filter beds. Pressure should be such as not to disturb the sand beds. Normal rates of application is 600 lpm/m^2, equivalent to a rise of 60 cm/ min for a period of 10 minutes. Air wash system is used to precede water wash of the beds. The combined systems of washing reduces the use of wash water, therefore, very useful in case of large filter units. Air is pushed through the underdrains for about 3.5 minutes to agitate the sand beds before introducing wash water.

The disposal of the waste generated during water

treatment process is itself a major task for the sanitary engineers and chemists. The solid wastes, produced in the water treatment plants mainly consist of following composition:

(a) Sludge from flocculation, residual chemical coagulants plankton etc.

(b) Wastes from backwashing of filter beds.

(c) Wastes generated during ion-exchange water softening process.

Slow sound filter

The erection of this type of treatment plant involves initially very high cost and need a large area than that of rapid sand filter process. The effective size of sand ranges from 0.2 to 0.3 mm. This process is very efficient in removing bacteria, colour and odour, even without the use of any chemical e.g. alum, lime and bleaching powder. However, this process is not at all suitable for the water which has turbidity value greater than 50 ppm and involves a large number of people. This process in fact is meant for the rural area where electricity is a problem and the availability of land is not a limiting factor.

The comparative merits and demerits for the rapid and slow sand filters have, however, been shown in the following table for assessing their feasible at a glance.

Table 5

Comparison of the Rapid and Slow Sand Filter

Parameter	*Slow Sand Filter*	*Rapid Sand Filter*
Compactness	Requires large area	Small area requirement gives compactness.
Chemical treatment	Not generally required	Essential

Filter media	Fine sand, effective size 0.2 to 0.3 mm	Coarser sand effective size 0.45 to 0.75 mm.
Filtration rate	2 to 3 lpm/m^2	80 to 100 1pm/m^2 may be increased if necessary
Flexibility	Not suitable for variation of demand	Suitable
Period of run	1 to 3 months	1 to 3 days depending upon the quality of raw water
Supervision	Skilled supervison not necessary	Skilled supervision essential
Economy	Initial cost very high	Economical
Suitability	Not suitable for water with turbidity above 50ppm	Suitable for turbid and coloured water
Efficiency	Very efficient in removing bacteria. Chlorination not necessary	Bacteria removal 90-95% Colour and taste are also removed. Chlorination essential

Pressure filter

Pressure filter is a type of rapid sand filter in a closed container through which water is passed under pressure. It is not considered very reliable in removing bacteria. Therefore it is better used to soften and remove iron from groundwater. It has a wide use in clarifying softened water in industrial plants. Nevertheless pressure filter is used in small water works in our country. A battery of units are installed, as per requirement, all receiving raw or chemically treated water and discharging into a common header pipe or pipes. Vertical pressure filters are generally 16-96 inch in diameter, while horizontal units are obtainable in 7 to 8 ft. diameter length varies from 8 to 25 ft.

Disinfection

Treatment of water by coagulation, sedimentation and filtration renders water chemically and aesthetically acceptable. Although a large part of the bacterial load is reduced during such treatment even then the water does not become absolutely free of them. Water, therefore, requires disinfection which implies the destruction of pathogen, i.e., (disease producing) bacteria.

Disinfection does not completely destroy all micro-organisms, which can be accomplished only by sterilisation. Complete sterilisation however, is not easily obtained, nor it is necessary, for making water potable. The commonly used chemicals and disinfectants are bleaching powder, liquid chlorine iodine, etc. The later is available in the form of tablets, known as tetraglycine hydro-periodide, galabaline, etc., iodine is used in disinfecting the water in such places, e.g., in the field of war, where no organised water supply facility exists.

(i) Chlorination

Chlorine not only removes bacteria but also eliminates taste, colour, and odour controls algae, protozoa and fungi, improves coagulation and oxidises iron and manganese. It is available as calcium hypochlorite in form of bleaching powder containing 33-35 per cent available chlorine (IS: 1065.1971) and in the form of HTH (High test hypochlorite) powder which is stable carries 65-75 per cent of available chlorine. These are used in small water works. For medium and large water works, chlorine gas is used because it is more economical and effective. It is compressed into liquid form and supplied in steel cylinders and is injected in gaseous state through chlorinators.

Chlorine reacts with water to form hypochlorous

acid (HOCl) and hydrochloric acid (HCl). The reaction is represented as:

$$H_2O + Cl_2 = HOCl + H^+ + Cl —$$

Hypochlorous acid being unstable dissociates into hydrogen ions (H^+) and hypochlorite ions (OCl) pt.

$$HOCl = H^+ + OCl$$

Both the above equations are reversible, the undissociated HOCl is 80 to 100 times more potent as a disinfectant than the OCl ion. The amount of dissociation is directly dependent upon the pH value of water. The disinfecting action is rapid when the pH value is about 7 or slightly higher. Beyond, it is reduced gradually because more and more HOCl dissociates into OCl ions. Equal amount of HOCl and OCl is present at pH 7.5.

Chlorine present in water as hypochlorous acid, hypochlorite ions and molecular chlorine is defined as free available chlorine. Normal chlorine does varies from trace to 1 ppm depending on the pollution load. The dose is determined with precision because disinfection will not be complete with insufficient dose, whereas a higher dose will form bad taste and odour, necessitating dechlorination. An effective dose of chlorine will leave a residual chlorine charge of 0.1 to 0.2 ppm which indicates that the water has been properly disinfected.

Chlorine, when added to water as free chlorine, oxidises some of the organic and inorganic compounds. While reacting with ammonia and organic amines, it forms chloramines. Chlorine present in water in chemical combination with ammonia or nitrogenous compounds is known as combined available chlorine.

Chloramines are also oxidising agents but are less potent than hypochlorous acid. The following reactions mark the formation of the three chloramines:

$$NH_3 + HOCl = H_2O + NH_2Cl \text{ (Monochloramine)}$$

$$NH_2Cl + HOCl = H_2O + NHCl_2 \text{ (Dichloramine)}$$

$$NHCl_2 + HOCl = H_2O + NCl_3 \text{ (Trichloramine)}$$

Monochloramine predominates at pH over 7.5 dichloramine at pH between 5 to 6.5 and odorous trichloramine (nitrogen trichloride) is produced at pH below 4.4. It is, however, volatile and disappears within a few hours retention period, but is carcinogenic in nature.

The chlorination of water to the extent that all the ammonia be converted into trichloramine (NCl_3)or oxidised to free nitrogen is termed as break-point chlorination. This is estimated by knowing the sudden fall in residual chlorine (RC) value in spite of continued chlorine addition to the same volume of water under test. Similarly, the difference between the amount of chlorine applied to the constant volume of water (say of one litre) and the amount of free, combined or total available residual chlorine remaining at the end of the desired contact period is known as chlorine demand of water. The latter determines the quality of raw water and thus, bears the positive relationship with the amount of oxidisable organic pollution load present in water. Raw water with a chlorine demand upto a value of 1.1 ppm is considered to be fit to be used as source of public water supply.

To ensure the wholesomeness of the drinking water, accurate estimation of free, residual, combined residual and total residual chlorine is essential.

The instrument used for chlorine estimation on spot is of pocket size known as chloroscope. Its manufacture has recently been commercialized in India and it is easily available from any scientific instrument dealer.

(ii) Methods for Estimation of Available Chlorine in Water

Put 0.5 ml of 0.2 per cent orthotoluidine solution + 10 ml water sample + 0.5 ml of 0.5 per cent solution of sodium arsenate ($NaASO_3$).

(a) Read immediately and discard-A

(b) Put 0.5 ml $NaASO_3$ + 10 ml after sample + 0.5 ml orthotoluidine solution.

Read immediately B_1.

Read after 5 minutes B_2.

(c) 0.5 ml orthotoluidine + 10 ml water sample, Read immediately C

Total residual chlorine = $C.B_t$,

Free residual chlorine = $A.B_1$,

Combined residual = $(C-B_2)$ - $(A-B_1$

At 20-25°C and pH 6.8, 0.2 ppm of free residual chlorine (FRC) for 10 minutes contact period is considered as the optimum dose for killing most of the bacteria or rendering them ineffective. The optimum FRC concentration (0.2 ppm) at the furthest points of the distribution network can only be maintained, provided, the following three points are satisfied.

(a) FRC should range from 0.5 to 1.2 ppm at the source.

(b) The distance between the source and the furthest distribution must be preferably within 10 kms, measured along the delivery pipes. For longer distribution system, re-chlorination will have to be done at suitable point to maintain requisite FRC.

(c) The minimum pressure 1.4 kg/cm^2 must be maintained throughout the distribution system, otherwise back flows of polluted water may occur.

(iii) Estimation of available chlorine in bleaching powder

Prepare 1 per cent aqueous solution of bleaching

powder. Out of the total prepared solution, 10 ml is taken out and to which 5 ml acetic alongwith 2 g solid potassium iodine are added. It is then titrated against N/40 sodium thiosulphate ($Na_2S_4O_3$) solution, using starch as indicator, till the end point comes. The calculation for the available chlorine is done as indicated below:

$$\% \text{ Available chlorine} = \text{ml of N/40 } Na_2S_4O_3 \times 0.89.$$

(iv) Other methods of disinfection

Boiling the easiest method to destroy all kinds of bacteria, alage etc. It needs vigorous boiling for 15 to 20 minutes. It is applicable for individual use because large volumes of water cannot be treated by this method. During epidemic, water must be consumed after boiling, even if the same is chlorinated, to become doubly sure about its sterility. Ultraviolet Irradiation is the most modern of sterilisation and consists of passing ultra violet rays from mercury vapour arc lamps. The astonishing results of this method is shown here.

Table 6

Killing time for Pathogenic Bacteria

Pathogens	*Time (sec.0)*
Cholera	10—15
Dysentry	13—20
Tetanus	20—60
Staphylococcus	5—10

Some special treatment for ground water

As has earlier been said, groundwaters are usually more hard as compared to the surface waters. Besides, the groundwater in some regions contains large proportion of dissolved elements viz. Iron, manganese, flouride, chloride, iodine, etc, which necessitate removal.

The groundwater in the extreme southern parts along the plateau regions across the border of Orissa, Madhya Pradesh and Bihar is acidic and as such need to be treated.

Table 7

Classification of Waters on the basis of their Hardness Value

ppm	*Degree of Hardness*
15	Extremely soft
30	Very soft
45	Soft
90	Moderately soft
110	Moderately hard
130	Hard
170	Very hard
230	Excessively hard
250 and above	Too hard for use

Hardness

There are two types of hardness, temporary and permanent. Temporary hardness is caused due to the presence of carbonate and bi-carbonate of calcium, magnesium and other bivalent metals. It can be removed by boiling. This is known as carbonate hardness. Permanent hardness is caused due to the presence of sulphates and chlorides of hardness constituting cations. This is also known as non-carbonate hardness and cannot be removed by easy means like boiling. These salts are very corrosive and deposit scales in steam boilers and in distribution system.

Waters having hardness values of less than 45 ppm is known to be very soft water and with more than 230 ppm as excessively hard water. The consumption of very soft water, especially in old age is desirable, because the toxic elements are removed from the body. The use of very hard water causes the loss of the tenderness of vegetables and meat cooked. Possible inflamation of intestines have been reported to be lessened or removed by consuming moderately soft water.

Removal of temporary hardness

(i) *Boiling*: Soluble Ca and Mg bio-carbonates are converted into their insoluble carbonates. The mechanism of the reactions is as shown below:

$Ca\ (HCO_3)_2 = CaCO_3{}^- + H_2O + CO_2$

$Mg\ (HCO_3\ )_2 = MgCO_3{}^- + H_2O + CO_2$

(ii) *Clark's Process:* Water is softened by the addition of necessary amount of lime in the form of lime water of milk of lime.

$Ca(HCO_3)_2$ is precipitated as $CaCO_3$. In case of $Mg(HCO_3)_2$ lime is added in double the quantity to form sparingly soluble $Mg(OH)_2$

$$Ca(HCO_3)_2 + Ca(OH)_2 = 2CaCO_3 \downarrow + 2H_2O$$

$$Mg(HCO_3)_2 + 2Ca(OH)_2 = 2CaCO_3 \downarrow + Mg(OH)_2 + 2H_2O$$

Lime softened water is not stable and causes deposits of calcium carbonate in pipes and filter sands. Addition of alum stabilises it.

Removal of permanent hardness

(i) *Soda-Lime Process* - Permanently hard water can be softened by the addition of sodium carbonate (washing soda). It converts the soluble sulphates and chlorides of Ca and Mg into insoluble carbonates of Ca and Mg. Water cannot be softened by this process below the hardness value of 21 ppm.

$$CaCl_2 + Na_2CO_3 = CaCO_3 \downarrow + 2NaCl$$

$$MgSO_4 + Na_2CO_3 = MgCO_3 \downarrow + Na_2SO_4$$

(ii) *Ion-Exchange Method*—Water of zero hardness value can easily be produced by this process Zeolites are used as the softener which are complex compounds of sodium, aluminium and silica. These are either artificially prepared or may be obtained from natural deposits and refined. They can easily exchange bases. The principle behind the method is that when hard water (both temporary and permanent) is filtered through zeolite, the calcium and magnesium cations are removed in exchange of sodium. The filtrate being free from Ca and Mg salts is soft. When the sodium of the zeolite is exhausted, it is regenerated by applying salt solution whereby an exchange is again affected. Softeners of the type are usually constructed like pressure filters.

System of distribution

Distribution of water is done either by intermittent or continuous system. In the intermittent system, wastage of water may be controlled but the main disadvantage is that during non supply hours, water in the pipeline rushes towards the lower reaches, thus creating partial vacuum in pipes, resulting in the sucking of impurities and gases in the pipeline. This may cause health hazards. Also during times of fire, intermittent supply system proved to be a great handicap.

Continuous supply, though preferable in everyway, but cannot be functional in every town, unless supply to every household is metered to avoid wastage. Grid system of distribution mains where all the pipelines are interconnected is always preferred. It eliminates dead ends.

Zoning

Zoning to the distribution is done for equitable distribution of water throughout the area. It mainly depends on the topography, density of population and type of locality.

Service reservoirs

It is of importance for regulation for pressure in the distribution system. It also regulates the supply in case of failure of electricity, pumps or any other such defects. The elevation of the reservoir should be designed to maintain minimum residual pressure in the distribution mains. In towns of undulating topography as well as in big towns, several service reservoirs are constructed at suitable places in different zones.

Pipe size and residual pressure

For towns having populating up to 50,000 minimum pipe size shall be 100 mm, whereas for population above 50,000 it should be 150 mm.

Minimum residual pressure at ferrule points in a distribution system shall be:

For single storeyed building	7 m
For two storeyed building	12 m
For three storeyed building	17 m

For higher buildings, boosting of water should be done by booster pumps.

Laying and maintenance of mains

For roads wider than 25 meters, distribution pipe shall be laid on both sides provided with suitable valves. Sluice valves should be provided at regular intervals and at junctions to isolate the portions for repair, when necessary. Air valves are provided to release air automatically when pipelines are filled. They are provided at the peak points in the pipeline gradient to release entrapped air. Absence of air valves may result in deficiency in flow or even bursting of pipes. Scour valves are provided at the valley points. They are opened at regular intervals to cleanse the pipeline while water reshes out to empty the pipeline for repair for replacement.

The use of hemp yarn while caulking with lead of cast iron pipe sometimes infects the water. The usual practice is to use bituminous coating as pipe lining materials is identified to increase the concentration of Poly nuclear aromatic hydrocarbon (PAH). Now-a-days, cement lining inside and zinc coating outside, of the cast iron pipe is gaining importance for smooth surface and longer working life. The latter is high accumulating in nature and known to be teratogenic (a substance which induces a change in the genes). The use of Poly Vinyl chloride (PVC) pipe in public water supply is also not good, because it contributes lead by leaching to the water if not properly controlled. PVC is not good for hot water supply.

Water supply to any individual building is made through house service connection. It is guided by the bye-laws of the concerned municipality.

Wastage of water

Wastage may occur in treatment plant, from leakages in service reservoir and pipelines, careless use by consumers through continuous flow in public hydrants, due to theft of cocks and so on. Levels of wastage upto 10% may be considered low but above it corrective measures are needed from the point of view of economy.

Protection against pollution near sewers and drains

Water main should be at least 3 m away from the existing or proposed sewer or drain. In case where it has to cross huse or sanitary sewer of storm drain the bottom of the main should be at least 0.5 m above the top of the drain or sewer.

Incrustation and corrosion of delivery pipes

Incrustation and corrosion are the major enemies of distribution pipes and valves, causing tremendous national loss in term of money and loss of out put. Incrustation significantly reduces carrying capacity of the water delivery pipes. Considerable amount of energy is lost due to incrustation in boilers. Similarly the aesthetic value of drinking water is tremendously lost due to incrustation and corrosion of cast iron pipes. The life of shallow hand pump pipes is drastically reduced from average length of 30 years to less than 5 years in regions where water is either acidic or contains exceedingly high proportion of soluble iron. Keeping in view the tremendous loss, caused mainly by incrustation of corrosion, an urgent need to envisage a proper check and balance method has been felt throughout the world. A laboratory method to ascertain the nature of water, along with easily applicable methods for the control are briefly presented here.

Laboratory method to ascertain the corrosive and incrustating nature of water

$$\text{Langelier Index: LI} = \text{pH} - \text{pHs} \quad \text{....(1)}$$

Where, pH is actual pH of water and pHs = Saturated pH produced by adding calcium carbonate.

The Langelier inded (LI) with zero value indicates that water is neither corrosive nor incrusting in nature i.e. the water is chemically balanced. A positive value indicates the tendency of water toward incrusting. A negative value indicates tendency to eat up the metals i.e. water under examination is corrosive in nature. The value of the saturation pH in equation (1) may be determined as such.

$$pHs = 9.3 + (A + B) - (C + D) \qquad ...(2)$$

The co-efficients A, B,C and D indicate the following factors

(i) Factor A-TDS
(ii) Factor B Water temperature
(iii) Factor C-Calcium carbonate in ppm
(iv) Factor D-total range alkalinity in ppm as calcium carbonate.

Prevention of incrustation

Incrustation cannot be prevented entirely, but it can be delayed and kept in check by keeping the draw-down as small as possible, as follows:

(i) In order to reduce the head loss to a minimum the well must be developed properly so that aquifer losses are kept at its minimum.
(ii) A screen with large open area and fully penetrating the aquifer should be installed. As such, the high precipitation of iron salts and carbonate will be reduced.
(iii) The pumping rate should be reduced by increasing the pumping period. The required quantity of water may be obtained from several wells rather than pumping from a few large wells at excessively higher rates.
(iv) The screens should be cleaned periodically, at least once in a year, even if the discharge has not fallen-off.

In case of already incrustated wells where treatment is essential, the same may be achieved by using acid, chlorine or other effective agents, depending upon the nature of the incrustated materials. Hydrochloric and sulphuric acids are effective in removing carbonates. Sodium hexa meta phosphate is effective in dispersing iron and manganese oxides. Chlorine is effective in removing bacterial growth and slimes.

Corrosion

The negative Langelier index results in corrosion which in turn, reduces the strength of the pipe, increases the screen size, cáusing sand pumping and eventual failure of the tubewell. The important reasons for corrosion are galvanic action corrosive water, high velocity and unsaturated waters.

(i) *Electrolytic or Galvanic Action* - It is the most important factor leading to corrosion. The principle behind electrolytic corrosion is that when two or more dissimilar metals are dipped into solution (viz. in water) a solution potential is set up which causes the electrons to move from cathode (electrode) to anode (electrode). Thus the anodic portion is corroded and deposition takes place on the cathode. The elements lying above the hydrogen in electrochemical series, act as anode and those below as cathode. The positions of the elements in electrochemical series determines the rate of corrosion, depending upon their mutual positions. General, the greater the difference in their position, more intense corrosion is produced. For example, if iron and brass are coupled together in water, iron acts as anode and brass as cathode. Thereby, iron is progressively being corroded and brass remains protected. Sometimes, corrosion takes place in the same metal between two path of slightly dissimilar composition. This is known as local corrosion. The protection of iron pipe is achieved by galvanizing the same. In doing so, the zinc start working as anode and iron as cathode. Thus, zinc is corroded and iron is protected from corrosion.

(ii) *Corrosive water* - If the water is acidic, it attacks the metals directly and causes corrosion. Hydrogen sulphide corrodes steel and copper alloys, if it is present in quantities greater than 1 ppm. If the dissolved carbon dioxide concentration is more than 53 ppm and DO is 2 ppm make the water corrosive. The electical conductivity value above 1500 mhos/cm makes water corrosive.

(iii) *High velocity of water* - The high flow of water removes the hydrogen from the cathode, which in trun accentuates the process of corrosion. Very low velocity of water tends to prevent local corrosion but makes it uniform.

(iv) *Unsaturated water* - When water is deprived of the minimum quantity of iron that a water can normally contain. It is known as unsaturated water. As such, the water corrodes iron and steel until it becomes saturated. The water with less than 50 ppm of hardness is generally unsaturated and often causes corrosion.

Methods to control corrosion

(i) The use of corrosion resistant materials can effectively control corrosion. Such materials include everdurbronge (16% copper, 3% silicon, 1% manganese) stainless steel, polythene, epoxy-bonded fibre glass etc.

(ii) Galvanzing. coating with plastics etc. offer considerable protection from corrosion.

(iii) Reduction in velocity reduces the rate of corrosion. It can be achieved by increasing the percentage of open area of the diameter of the well pipe or decreasing the rate of pumping or the draw down.

(iv) The corrosiveness of water may be affectively reduced by suitably treating such water. The addition of lime to neutralise the acidity of water is worth suggesting.

(v) Cathode protection: In this method an alternative anode is selected. Thus, the iron pipe may be protected from corrosion by making zinc as anode.

The water is generally made to chlorinate at the end of treatment with enough chlorine to kill essentially all bacteria and viruses and leave a residue of chlorine sufficient to keep the water safe until it reaches the water tap. As water being treated for drinking purposes often is contaminated by organic substances, chlorination can produce some organic compounds of chlorine. Some that have been detected have been chloroform and carbon tetrachloride. These compounds are suspected of being carcinogens.

No alternative technology to chlorination is immediately available. Ozone is used by special permission in one very small system. Ozone has been highly toxic to bacteria, and it destroys viruses more efficiently than chlorine. It leaves no taste. The technology for the use of ozone has been somewhat more expensive than for chlorination, and we have slight long-term experience in the use and effectiveness of ozone. Thus the government must move very carefully before requiring a switch from chlorination. Eventually the change could be made. The methods of purifying waste domestic and industrial waters are largely traditional and have proved extremely effective in the past. However, in recent years further methods of purification have been introduced, particularly in the field of industrial water treatment.

Use of coagulants

For effective removal of impurities from water it becomes essential that these have first flocculated, that, have been in a form that could be removed readily by filtration method. The aim has been to obtain particles in excess of 120 microns in diameter; these tend to settle readily in typical aqueous medium. Particles between about 3 and 100 microns tend to settle too slowly for practical purposes.

Flocculation seeks to achieve agglomeration of small particles into larger units and thereby speed up the settling process. The most widely used flocculating agent has been alum, a complex salt of aluminium. When this has been added to an alkaline solution a voluminous aluminium hydroxide flocculate has been produced, which readily sinks to the bottom because it has been heavier than water. In its passage downwards it triaps tiny particles suspended in the water and thus removes them.

A recent development has been the use of a number synthetic and natural and polymers. These form 'bridges' between adjacent solid particles and cause them to sediment. In many instances polymer sedimentation has been not as effective as alum, because small particles, particularly those with a diameter below 2 microns, tend to possess a (usually negative) electric charge. Alum is able to dissipate this charge through the presence of positively charged aluminium ions in solution. However, metal salts have been not without their disadvantages. They are expensive, sometimes give inadequate flocculation and may also leave a 'carry-over' of fine turbidity.

Special synthetic polymers are being manufactured which are having the properties of bridging and discharge of electrical potential. Only water soluble monomer polymer systems can be used for waste water clarification. The following poly-electrolytes have been used.

(a) *Non-ionic types.* Polyols, polyethers, polyamides, polyvinyl heterocyclics;

(b) *Anionic groups.* Carboxylates, sulphonates, phosphonates;

(c) *Cationic groups.* Amines, quaternary ammonium compounds, sulphonium and phosphonium compounds.

The most common of all flocculants have been the polyacrylamides which are having molecular weights in the range 4-10 million.

Flash distillation

Pure evaporation of seawater and other brackish water has been still the most widely used of all methods to recover the pure water. The heat needed to boil a kilogram (litre) of pure water at 100°C has been nearly 2.27MJ; but the heat needed to boil concentrated solutions of salts is greater as the presence of the salts elevates the boiling point. The main economy of the process has been dependent upon obtaining adequate heat economy.

This could be mainly achieved in two ways. First, solutions can be evaporated at lower pressure when the latent heat needed has been markedly reduced; and second, attempts can be made to use low grade waste heat for the process, especially using waste fuels like refuse, petroleum tail gases, etc., for operating seawater distillation plants.

There has been a considerable future in the operation of seawater desalination plants which use distillation on the total energy principle. The most commonly used process employing distillation as a basis has been flash distillation.

Flash distillation is involving a technique of allowing the seawater to boil successively in a large number of chambers, each of which has been maintained at a lower pressure than the one before, to match the lower temperature of the water. The method of flash distillation of seawater has been especially promising when related to the use of waste heat produced in nuclear power stations. It has been feasible to design a nuclear power stations; operating on the total energy principle, to produce 400 MW of electricity and 283,000 m^3 of freshwater per day.

Operating cycles of a flash evaporator. The flash evaporator is having twenty to fifty chambers in sequences, which are operating at successive lower pressure. When heated brine flows from one chamber to the next, some of it 'flashes' off into water vapour, which then condenses on colder condenser tubes and drops as distillate into trays to be led away into storage.

The brine, when passing from chamber to chamber, gets progressively cooler and it has been this same brine which is eventually pumped back through the condenser tubes to act as the coolant in the condenser section of each chamber. It becomes progressively hotter as it does this; consequently, when it reaches the heat input section before re-entering the first flash chamber, its temperature need only be raised a few degrees. The heat would be normally supplied by low-pressure steam which, in its turn, has been readily obtained by utilization of waste heat from primary power generating sources.

The heated brine has been passing from the heat input section to the first flash chamber (1) and from there successively through all the chambers down to the coolest one (6), flashing off a certain amount of water vapour at each stage. It could be then extracted by the brine circulator pump and returned to the tubes of the heat recovery section at (4). In the last few stages-the heat rejection section (5) and (6)—cold crude sea water has been pumped through the tubes.

This allows condensation to occur in these stages and also extracts an amount of heat equal to that put into the evaporator at the heat input section, thereby allowing permitting a continuous cycle of operation. Some of the seawater, after chemical treatment to disallow scale formation, is added to the circulating brine to make up for the distillate extracted and for brine which must be discarded in order to keep the solute concentration in the evaporator within specific limits.

When seawater has been heated gases such as oxygen and carbon dioxide are given off. These could be extracted by a vacuum pump or ejector system.

The heat needed in a flash evaporator to produce unit weight of distillate has been given by the following equation:

$$(t_e - t_1)/(t_e - t_f) = L$$

where t_e represents the temperature of the water enter in the flash stage in °C;

t_1 represents the temperature of the water leaving the tube system in °C;

t_f represents the temperature of the water leaving the first flash stage in °C;

L represents the latent heat of flashed vapour in KJ/ kg.

Capital costs for typical evaporator plants have been found to depend upon the operating efficiency. Low efficiency plants are having a $(t_e - t_1)/(t_e - t_f)$ ratio of 0.25, while the ratio for high-efficiency plants has been of the order of 0.1 or less. Thus low efficiency plants are having a fuel consumption which has been 2 1/2 times as high as that of high efficiency plants.

Solar stills

These methods have been likely to gain importance in area of the world where fuel is scarce but there is adequate heat from the sun. Solar radiation energy available in many parts of India and Pakistan has been of the order of 20MJ per m^2 land surface area per day for some five months in the year, and even in mid-winter the solar heat has been still more than 10MJ/m per day.

A system developed in India uses a large area of glass positioned between 30 and 60 cm above the surface painted black, along which seawater is allowed to flow. Solar heat makes water to evaporate, leaving the salt behind, and condensation takes place on the glass cover. The water drips off to an aluminium channel and runs to a strange vessel. The unit has been also equipped to collect rainwater. The cost of getting water from the experimental $2m^3$ per day stills, which have been able to extract around 3-4 litres of distilled water from each m of still surface erected, is about 18p per m^3. It is now thought that with larger installations it becomes possible to use this system anywhere the present cost of water exceeds 70p per 1000 US, gallons or 18.5 per m^3. Such conditions have been found in many parts of India and other arid regions of the earth.

A large solar distillation plant was built near Daytona Beach, Florida. The feed water is having 3.2 per cent dissolved salts and the area of land covered is $220m^2$. Each unit is consisting of grooved concrete curbs laid 1 m apart to form a bay 20 m in length.

There have been plastic basin liners and clear plastic cover domes. Underneath the brine has been a thick layer of insulation, so that the heat generated inside the still has been not conducted away to the soil. Water evaporates from the feed water and condenses against the plastic dome cover, running down towards galvanized steel distillate throughs at the sides of the stills. Solar stills, based on similar designs to the above, are also erected in Greece and Spain.

For a basin type solar distillation plant, with a capacity of 180 m^3 of water per day, covering a ground area of 50,000 m^3 plant investment cost has been about 2,00,000; as far as running costs are concerned, amortization accounts for 70 per cent of the daily costs.

The next major items of expenditure have been taxes, insurance and interest on working capital, which add a further 18 per cent. The final cost of water supplied works out at 35p per m^4 very much higher than with distillation systems which use purchased fuels.

However anomalous it may be that where the fuel has been free the costs has been higher, it has been certain that solar stills will not be economic in comparison with other methods of water desalination for many years to come.

Ion exchange

The cations commonly present in natural waters including seawater, have been calcium, sodium, potassium, magnesium, iron, and traces of other metals like manganese, aluminium, copper and zinc. The usual anions have been bicarbonate, chloride, sulphate and traces of nitrate and fluoride.

Ion exchange purification takes place by passing water first though an exchanger where cations get exchanged for hydrogen ions, followed by passage through an anion exchanger bed where acid radicals have been replaced by hydroxyl ions so that:

H^+ (hydrogen ions) + OH^- (hydroxylions) $\rightarrow$ H_2O (Water)

Naturally, it becomes necessary to reclaim both the

cation and the anion exchange resins as they become spent. This is normally carried out by reacting the former with strong acid and the latter with caustic soda. There have been a number of more sophisticated plants including so-called 'mixed bed' de-ionizers where special methods of fluidization are able to separate the different ion exchange resins to enable them to be reclaimed separately. In the past it was regarded uneconomic to use ion-exchange plants for the desalination of water with a solids content in excess of 1.5 per litre.

By employing a cascade system of desalination it now becomes possible to treat water with a salt content well in excess of 10 g per litre. Some Russian plants have been using ion exchange membranes for the desalination of waters having upto 15 g of salt per litre. The technique used has been a combination of electrodialysis, and ion exchange using membranes which are referred to simply as MK-40 and MA-40. These are having as surface electrical resistance of about 12 ohm/cm^2, measure 150 cmx50 cm have been able to exchange between 2.3 and 4.4 mg equivalent of ions per hour. The total electric power consumption needed has been 9 kWh per m^3 of drinking water produced.

A new American technique employs weak electrolyte ion exchangers to change sodium chloride into sodium bicarbonate first followed by removal of the cation (sodium) and then recovery of the carbon dioxide which is still present in solution. It has been reported that with this technique really concentrated solutions of salt could be treated efficiently and economically. The 'Sirotherm' process has been an ion-exchange process in which regeneration of the resin has been by heat and not chemicals. At present his process has been still at the laboratory stage, but there have been indications that it has been technically, feasible to treat water with a salinity of upto 3 per litre. It is thought that some other process may be required to reduce the salt content of seawater from its usual 35g per litre to that at which an ion exchange process is feasible. This implies a combination of electrodialysis or reverse osmosis with an inexpensive ion-exchange technique like 'Sirotherm'.

Reverse osmosis

When we keep a solution on one side of a semi-permeable membrane and pure solvent on the other, the solvent will pass through this membrane to dilute the solution. The pressure which gets exerted by the solvent as it passes through is called the osmotic pressure.

Indeed, it is to be argued that in solution a dissolved salt behaves like a gas and obeys the gas laws. This implies that one gramme mole of cation or anion dissolved in 22.141 litres of water gives an osmotic pressure of exactly 1.013 bars (1 atmosphere) at a temperature of 0°C. Reverse osmosis has been found to depend upon the use of a semi-permeable membrane, which is capable of being traversed only by water and not by salt, combined with pressurizing of the solution so that water passes through the membrane, but the salt stays behind.

Due to the high pressures involved, the main difficulty has been in the making of a semi-permeable membrane of adequate strength. In general, membranes could be supported by a special carrier. The following semi-permeable membrane materials are tried, with favourable results:

(a) Cellulose acetate dissolved in glacial acetic acid together having small quantities of additive like magnesium perchlorate before being allowed to solidify on the carrier;

(b) A cellulose acetate/formamide solution in acetone;

(c) Ethyl cellulose mixes.

The working pressure on one side of the reverse osmosis cell has been normally of the order of 100 bars, while the other has been at atmospheric pressure. Supporting the semi-permeable membrane to avoid breakdown has been therefore a problem of considerable magnitude.

The first reverse osmosis plants was using perforated stainless steel plates and tubes as supports, but these have been reported to be too expensive for commercial plants. The best material found which was having the requisite strength and at the same time was not too expensive was fibreglass cloth laminate about 3 mm thick; this is having a compressive strength of 15 MN per m^2 and a tensile strength in excess of 150 MN/m^2. As far as tubular support materials are concerned, phenolic resins having a wall thickness of 2.5 mm seemed to provide some of the best results at the lowest cost.

The desalination cell has been so designed as to bring the saline solution at high pressure in contact with the semi-permeable membrane, with either plates or tubes to support the membrane itself. As osmotic pressure tends to increase with increasing salinity, the through put rate for a given membrane is less with seawater (3.5 per cent dissolved solids) then typical brackish water (less than 0.5 per cent dissolved solids). The final product usually is having about 2000 ppm of dissolved solids, and may have to be purified further by other means.

Reverse osmosis has been found to be more economical for low salt content water than for seawater. As the osmotic pressure of seawater has been approximately 25 bars, it has been obviously necessary to overcome this pressure before reverse osmosis can begin.

Osmotic pressure builds up as the feed stream is concentrated and this very much is able to limit the percentage of water in the feed which can be usefully recovered as freshwater; only about a third of the water can, in fact, be recovered. Recovery of some of the power from the flow of the waste stream of concentrate seawater has been usually attempted.

It is found that a reverse osmosis plant producing 40 m^3 of water from seawater would cost about 3.5 million, and its operating cost would be 4.5 p per m^3 of water obtained. A plant to recover pure water from brackish sources would be costing about £1.1 million with an operating cost of only 2.3 p per m^3.

In spite of the disadvantages of reverse osmosis, this system has been likely to become one of the most important in the field of water desalination.

Practical reverse osmosis plants. Unlike many of the other desalination processes, reverse osmosis has emerged from the pilot plant stage and has been now used industrially. Paterson Candy International make reverse osmosis plants able to supply drinking water from brackish sources to private houses, hotels, farms, small industrial undertakings, etc.

The units are using a feed of brackish water with a maximum dissolved solids content of about 1 per cent and pH between 3 and 7.5. The salt reduction achieved has been between 85 and 95 per cent, depending upon the nature of the ions. The semi-permeable membrane in Paterson Candy units has been modified cellulose acetate cast on the inside of disposable paper/polyster tubes.

Normally membranes can find use continuously for between 12 and 24 months before replacement becomes necessary. An electric motor has been used to produce a pressure of 40-80 bars, depending upon the nature of the feed. Power consumed is, on average, about 2.5 kWh/m^3 for distillation water—which compares favourably with 12 kWh/m^3 for distillation and 7 kWh/m^3 for freezing processes.

A plant measuring 2.6 + 1.5 x 1.3 m and weighing under 2 tonnes has been able to produce 45 m^3 of desalinated water per day. This has been incidentally, the largest plant currently made by this firm, which produces equipment having capacities from 2.3 to 45 m^3/day.

Electrodialysis

In electrodialysis, as in reverse osmosis, a semi-permeable membrane permits pure water to pass, but stops salts. Whereas in reverse osmosis the passage of water through the membrane has been induced by physical pressure acting on the saline solution, in electrodialysis this work has been carried out by an electric current.

The basic idea has been that the saline solution has been enclosed by membranes and an electric current induces cations (metals) to migrate outward towards a cathode, and anions (the non-metallic ions) to migrate towards an anode, thereby progressively desalinating the solution.

Unfortunately, the procedure in practice has been not quite as simple as this. For progressive desalination to continue, the energy exerted during the electrodialytic process must also overcome osmotic pressure which is able to force clean water back into the solution.

As in reverse osmosis, membranes have to be very tough indeed for withstanding the high pressures applied. At the present time electrodialysis has been most widely used in the Soviet Union, and the most common membrane has been polyethylene backed by nylon fibre.

In general the greater the current density used in any given dialysis cell, the smaller the area of membrane required. But increased current density also necessitates increased flow rate to prevent the membrane from being polarized. Optimum current density has been found to vary with the concentration of salts in water to be treated, the cost of the membrane and the cost of the power expended.

For seawater having a salt content of 3.5 per cent and a power cost per kWh amounting to 0.14 per cent of the cost of membrane per m^2, the optimum current density has been 1.06 A/dm^2. For other conditions, the optimum current density may decrease between 0.7 and 3 A/dm^2.

Electrodialysis desalination finds use in the U.S.S.R. mainly for water supply in sparsely polluted areas, farms and watering places at distant pastures—as well as for expeditions venturing to areas where there has been no fresh water but there has been access to underground or surface salt water. Although the process has been basically an expensive one - both from the point of view of initial cost and of power needed - it is having the considerable advantage that it can be completely automated, needing no resident labour.

In Western countries electrodialysis for seawater desalination has been not generally regarded economically viable compared with other methods, mainly due to the large quantities of power needed. However, if at some future date the cost of electric power must fall appreciably, electrodialysis may become competitive.

Separators of expanded rigid PVC have to be used to support the membranes and to create turbulence in the stream. Two plants have been built, one with a daily output of 77 m^3 and the other with a daily output of 151 m^3, using membrane areas of 12 and 40 m^2 respectively.

The biggest American electrodialysis plant is the 2460 m^3 per day plant at Buckeye, Arizona. This brings down the salinity of the water from approximately 2.1 per cent to 0.5 per cent at a cost of about 5 p per m^3 of water treated. Electrodialysis provides the most economic results when the installed plant capacity is large and when the feedwater salinity is comparatively low.

Ion Exchange Techniques

There are four basic ion exchange resins which have been able to absorb even small quantities of dissolved cations from solutions. These have been sulphonic-phenolic resins, sulphonated coal, carboxylic resins and sulphonated polystyrene. For all of them the reaction which occurs can be expressed as:

$$HR + M^{+} \rightarrow MR + H^{+}$$

where M^+ represents the metallic ion (its valency does not have to be one) and R is the net-like resin structure. Once the ion exchange resin gets absorbed its capacity of metallic ions, it could be reclaimed by means of a strong acid:

$$MR + HCl \rightarrow HR + MCl$$

The metal can then be able to recover from the concentrated chloride or other salt. Ion exchange of this type has been found to be very effective for many metals especially, lead, copper, zinc, mercury, chromium and nickel. For arsenic an anion exchange resin is able to remove the metal in the form of arsenates and arsenites; reduction of arsenic contents of waste liquors to below 5 ppm can be attained without difficulty.

Copper and zinc can be removed from waste liquors

so effectively that a final effluent is having less than 0.001 ppm of either metal. It has been vitally necessary, as it has been shown that even concentrations of 0.02 ppm of these metals are having very harmful effects on fish. Ion exchange techniques must be used, where possible, in conjunction with precipitation methods, because they have been most successful when the initial concentration of dissolved metal ions has been already fairly low.

Self-purification of Waters

In the various zones of receiving streams (surface waters such as streams, rivers and lakes into which unpurified or purified waste waters are channeled and by which they are carried away), living creatures are to be found which are able to utilize any nutrients that are available. Natural self-purification of the waters by the decomposition of putrefactive substances, after a given period and distance of flow of the receiving streams, is due to these creatures. Under favorable conditions the pollution of the waters is substantially corrected in this way, so that the original pure state is at least approximately restored.

The self-purification processes depend mainly on biological processes (self-purification by organisms) or chemical processes (primarily oxidation and reduction processes), the effectiveness of which is determined by physical factors such as velocity of flow, state of the water bed, ratio of foul to clean water, temperature and depth of water, intensity and duration of solar radiation and fineness of waste substances. The main burden of self-purification falls on the biological processes.

The organisms contributing to the process of self-purification include fungi, algae, bacteria, crustaceans, shellfish, worms, insect larvae, snails fish, big predatory animals and waterfowl.

In the metabolism of the organisms these substances are incorporated in the body matter or ultimately broken down into water and carbon dioxide for the purpose of providing energy. In order to sustain these processes, dissolved oxygen is absorbed from the water. Owing to their autotrophic manner of feeding, the algae containing chlorophyll are in a position by day to return the oxygen so absorbed. If the water is overrich in nutrients, i.e., overloaded with plant nutrients such as nitrogen or phosphorus, there can well be an explosive proliferation of algae. This will result in a shortage of oxygen as, owing to the limited absorptive capacity depends largely on the water temperature. At higher temperature the saturation point is reached more rapidly. If algae are present in the water in over abundance, the oxygen requirements at night will lead to a severe oxygen deficiency.

The protozoa, which absorb bacteria and algae as nutrients, are also able to utilize dissolved organic substances to some extent. Crustaceans, snails and worms feed on undissolved, deposited or suspended matter. They also ingest protozoa. Bigger crustaceans and insect larvae, for their part, feed on small crustaceans, worms and also on protozoa. Fish and waterfowl, which are also liable to fall victims to predatory animals in their turn, often prey on crustaceans and worms.

The intensity of self-purification is determined not only by optimum functioning of the living communities but also by the conformation of the river-bed. In naturally formed river-beds that is to say, in waters with large surfaces and strong turbulence (irregular strong water currents of high turbulence), the higher input of oxygen makes for more favorable living conditions for the organisms than are found in corrected or dammed river courses.

Even minor changes in one or another of the determining biological, chemical or physical factors can completely upset the self-purification capacity of the waters.

Sewage purification

Until relatively recent times, the self-purification capacity of static and running water was still sufficient to reduce the contaminants contained in them. Loading the waters with waste and sewage from different sources has proliferated, however; and owing to the accretion of these contaminants, the reduction processes in the waterways have either come to standstill by the extermination of micro-organisms or been intensified by the increase in nutrients. With the multiplication of organisms, more oxygen has been extracted from the waters. The self-purification capacity of the waters has been overloaded, and the regenerative powers of the waters have been disrupted. The immediate consequences of such pollution have been the extermination of fish life, deposition of mud, and putrefaction processes.

The most serious loading of the waters stems from domestic and industrial waste disposal. The sewage is conducted through the drainage system to a purification plant, which as a rule operates in two stages: mechanical and biological treatment stages. The mechanical system comprises a grid, a sand trap and the preliminary settler. The coarser elements in the sewage are separated and retained by the grid. As the flow speed of the water slows, smaller mineral elements settle in the sand trap; the other removable matter is deposited in the preliminary settler, and the floating elements are removed by means of a scoop. The purification performance at the mechanical stage stands at about 20%-30%.

The water thus purified passes finally to the biological stage, which comprises a restoration basin (percolating filter) and a second filter, or final sedimentation tank. In the restoration basin, bacteria and other micro-organisms convert the suspended and dissolved matter into removable sludge by absorbing the particles and dissolved matter as a nutritive substratum and by forming cell lumps as they proliferate, which, owing to their increase in weight, sink to the bottom. The whole process is facilitated and intensified by artificial induction of air by means of bellows, rotors and cylinder pumps.

In order to ensure that the micro-organisms required for decomposition are present in sufficient quantities, parts of the micro-organism sludge deposited on the floor of the secondary filter tank are returned to the restoration basin. The purified water on the surface of the secondary filter tank is fed into the receiving stream.

The sludge from the primary and secondary settlers is digested in septic towers with the aid of anaerobic bacteria. By this process, methane gas is formed which can be used to provide energy for the plant. Previously the digested sludge was usually dried out in beds to compact it. In new plants the preference is for mechanical dehydration plants, such as centrifuges and presses, because of the smaller space they take up. The dehydrated sludge can then be further processed together with household refuse.

Under optimum conditions the mechanical-biological sludge drying plant can achieve a purification performance of over 90%. For further purposes, however, this performance level is still too low, so that a further stage, the chemical stage, needs to be added. The chemical precipitation process employed serves to eliminate phosphates and other (especially industrial) pollution. With the aid to microsieves and similar technical appliances, organic residues can be further reduced. In addition, a hygienic improvement in the sewage may be achieved by means of chlorination, radiation, heating or ozonizing.

The corrugated plant filter

For the purification of contaminated waste waters, a plant with corrugated-plate-filter much piles can be installed wherever the water is charged with floating or deposited matter. The oil filters can be installed at any point in the petrochemical or mineral oil processing industry, in tanker depots, at airports, in big garages, etc. Plant for the removal of solids is installed for the purification of filter return water and for sludge filtration (e.g., in the foodstuff and textile industries).

The corrugated plate filter depends for its operation on the force of gravity. In the filter the particles of contaminants do not need to rise the whole way to the surface of the basin, not to sink to its floor, since the filter contains a number of corrugated plates, one above the other, representing in effect a series of surfaces and of floors. Any one particle need to cover only the distance between two plates and is then effectively deposited. The corrugations on the plates make for closer pooling of the contaminants from the water for onward transmission to the filter channels. Additional collecting channels on the inlet and outlet sides of the corrugated plate will ensure that the filtered matter is collected and carried away without being diverted (by the inlet or outlet currents) into the sewage water again or contaminating the clean water.

According to the specific gravity of the contaminants to be eliminated, the filter will be fed from above downward (for matter that is lighter than after, such as oil) or from below upward (in the case of matter that is heavier than water).

The way in which the filter plant and the filter itself function is illustrated in the following example, relating to oil filtering. The inlet tank produces an even distribution of the incoming contaminated water between the individual parallel cells of the filter tank, so that all the plate packs are equally loaded. In front of each pack a flow filter spreads the water flow evenly over the whole field. The oil and water mixture makes its way from above into the pack, which is set at an angle of 45° to the vertical. It flows down diagonally between the corrugated plates set one above the other. The flow between the plates is laminar; the oil droplets rise unhindered to the underside of the plate above, collect in the crests of the plate, and flow back upward to the entrance of the plate pack. There the oil is caught up in the mainstream and carried away from the flow of water coming in to the surface of the basin.

The sludge present in the water collects in the troughs of the undulations, sinks down with the current flow of the water to the outlet of the pack, where it is similarly taken up by the collecting channel, and is carried in the mainstream to the floor of the basin or into an elutriating funnel. The cleansed water passes into the outlet tank, and from there over an overflow dam into the outlet channel. The separated oil collects as a slick on the surface of the water in the filter tank and flows over an oil scoop or skin pipe into the pump. The filter tank is covered with floating plates of plastic form in order to reduce evaporation and offensive fumes.

For the separation of solid matter the direction of flow is reversed, but the way in which the entire plant works is, in other respects, the same. The average performance of a set of plates, with a distance between the plates of 19 mm, is 30 cubic meters per hour. In special cases this average performance is reduced, however, to enable particles with lower rates of rise or fall to be thoroughly separated.

Composting of sewage sludge: The counterflow process

The counterflow process, which was developed for the composting of raw sludge from local sewage plants, is suitable for use in all sewage plant systems. The reactors are available in sizes of 25 cu.m, 50 cu.m, 100 cu.m, and upto 500 cu.m. Anaerobic or aerobic sludge stabilization (digestion tower) is superfluous. What is necessary, however, is drainage of the sewage sludge by mechanical means. For composting the sludge, owing to its high nitrogen and water content, an organic carbon carrier, such as peat, sawdust, chopped straw or lignite, is also required.

The sludge, fed into the sewage plant with a water content of about 98%, is dehydrated mechanically by means of presses and centrifugal driers down to a water content of 75%-80%. The dehydrated sludge can be mixed with the required organic carbon carrier by means of a mixing screw in the ratio of 1:1. In the case of a moving operation, instead of the carbon carrier, backflow (reflux) matter can be used in the proportions of 50% sludge, 40% reflux and 10% carbon carrier. The mixed material is raised by a sloping conveyor into the bioreactor, which consists of combined metal elements. In order to prevent heat dissipation, the reactor is insulated over its entire outer wall. The material provided passes slowly and continuously through the reactor from top to bottom (period of passage about 10 days). Ventilation is effected at the base of the reactor. The air injected through nozzles passes upward through the material. Control of the airflow is possible. Air is constantly removed at three points in the material store (reactor filling) and its carbon dioxide content examined with analysis equipment. Resistance pyrometers constantly check the temperature at six different points, which is then recorded on a six-point ink writer.

The rapid group system is adopted for utilizing clean sewage sludge. By a regular supply of oxygen, optimum living conditions are provided for the reducing and converting microorganisms already present in the sludge. In this way they are stimulated to exercise higher metabolic activity and faster germination. The air injected has an initial oxygen content of about 21%. It penetrates the material slowly by a counterflow process in an upward direction. This promotes the respiration of oxygen by the organisms. Oxidation process thus get under way and the oxygen content of the air is reduced from below upward. In the upper field of the reactor the content of oxygen in the air falls to 7%-10%. In accordance with the airflow the temperature rises from below upward. In the upper region immediately below the condenser water zone the temperature builds up to levels of 75°-82°C (167°-180°F). The material, which is slowly and continually traversed by the reactor from top to bottom, must necessarily therefore pass through this heat accumulation zone so that hygienic treatment of the sewage sludge is assured. The end product, considerably reduced in volume and weight compared with the original material, is a hygienically flawless, friable material.

Water supply preparation of surface water

Water consumption in big cities amounts to some 200-300 liters (210-315 quarts) per person per day, rising in peak periods to as high as 450 liters (472 quarts). On top of this come industrial water requirements. To produce 1 kg (2.2 lb) of plastics, upto 500 liters (525 quarts) of water may be needed, and to produce 1 kg of paper upto 3000 liters (3150 quarts). Of the total supply of drinking water today about half comes from groundwater, about one-third from spring water, and about 15% or 16% from surface water in lakes and rivers.

In thickly populated industrial centers, groundwater supplies as now utilized practically to the limits of their capacity. Increasing recourse must consequently be had in the future to the surface water in lakes and rivers. But because of the high cost of treating surface water for drinking, the supply of drinking water from this source to wider sectors of the population will raise the price of water considerably.

An example of a river water treatment plant will be described: The untreated water is drawn from 'the river by means of a suitable suction plant and conveyed to the water works by a remote-control pumping plant. There the river water is sterilized with chlorine in premixers and treated with a flocculation agent. This produces heavy sludge flakes which sink to the bottom. The clear water remaining above flows on to the ozonizing plant. There ozone gas is emitted into the water in fine bubbles, by which means (much more intensively than by chlorine) additional sterilization and oxidation of dissolved organic matter are achieved.

The ozone gas needed for this purpose is generated from oxygen at the waterworks by means of electric energy, and it reverts to oxygen again afterward, leaving no trace. In the adjoining rapid filter plant all the other still-dissolved foreign bodies are removed from the water, first of all the remaining suspended particles with the aid of a 2-3 meter thick quartz sand layer and then all the other still dissolved foreign bodies by means of the active carbon filter inserted below it. The treated water is retained until its introduction into the filtrate reservoir.

A further possibility is to allow the treated water to seep away into suitable gravel deposits in the river valleys, to lower its temperature and assimilate it in other respects to natural groundwater, and after a specified period to remove it in the same way as groundwater together with natural groundwater. A pure water reservoir is needed to retain the water removed, from which the drinking water is conveyed by pumps to the final consumer. Consumption of drinking water has risen in the past 50 years about 40 fold. This is explained, for the most part, by the enormous expansion in industrial production, by increased awareness of hygienic requirements, and by changes in man's consumption habits. Furthermore the population has practically doubled in the period.

The proportion of surface water in total supplies of drinking water is likely to expand in line with the constant development of industry.

Fluorine and the fluoridation of drinking water

Fluorine, a member of the halogen family of elements, is the most corrosive chemical element that is known. It is a greenish-yellow gas, which, like the fluorine compounds, is highly toxic. The addition of fluorides to drinking water to prevent dental caries (tooth decay) had been opposed by some doctors and toxicologists but now has been endorsed by all major U.S. and international health organizations. The application of fluorides in tooth-pastes or for internal consumption hardens the enamel on teeth and so provides protection against destruction of the teeth by caries.

Against the fluoridation of drinking water, various toxicological and legal arguments have been adduced. Water is necessary for life, whereas fluorine and its compounds can be toxic substances. It is for the water works to provide drinking water that satisfies the hygienic requirements of comestibles (food). Fluoride at high concentration produces a corrosive effect on glass, steel and a number of other metals. Drinking water is treated with 1 mg of fluoride upper liter of water. At concentrations over 2 mg per liter, people begin to react gradually with symptoms of mild dental fluorosis, as part of the daily intake of fluorine accumulates in the body. The fluorosis manifests itself as white flecks in the dental enamel. Pitting of the enamel can result from fluorine concentrations of 3-4 mg per liter. Only in a few places in the United States is the concentration above 4 mg per liter. In those areas the Environmental Protection Agency is working to reduce the fluorine level.

People do not absorb fluorine into their system only as an additive in drinking water, but fluorine is present in their diet in natural and artificial forms. This supply of fluorine in the daily diet varies in different countries and for different types of food, so that the total intake of fluorine in drinking water and food-stuffs cannot be assessed precisely.

A further factor is the loading of the atmosphere with the fluorine waste gases from aluminium works and similar industrial plants, which can also result in an uncontrollable intake of fluorine.

Although fluorine has been shown active in a dental context only for children under 15 years of age, the entire water supply system is treated with fluorides. It is estimated that about 1000 liters of water would have to be treated with fluoride to enable a child whose teeth are still developing to drink 1 liter of treated water.

A difficult legal problem now arises from the fact that the consumer can in no way avoid being supplied with fluoridated drinking water. The situation amounts to enforced medication. However, the legal validity of fluoridation has been upheld in courts in the United States and in several other countries.

In many cities in the world, however, the fluoridation of drinking water introduced in the past on a trial basis has been abandoned. In 1971 the model project introduced in 1952 in Kassel-Wahlershausen, Germany, was suspended. In 1972 the fluoridation of drinking water in Sweden was discontinued. In the United States more than half the population drinks fluoridated water and that proportion is increasingly slowly.

Caries is not a disease caused by fluorine deficiency but is the result of bacterial action exacerbated by general nutrient deficiencies and lack of dental care. Fluorine medicatioı can, at best, influence the symptom of caries, but not the disease itself.

The signs of water pollution have been obvious to even the most casual observer. To aid in a systematic discussion of water pollutants, they will be classified into 9 categories given below.

1. Oxygen-demanding wastes
2. Disease causing agents
3. Plant nutrients
4. Synthetic organic compounds
5. Oil
6. Inorganic chemicals and mineral substances
7. Sediments
8. Radioactive materials
9. Heat (Thermal Pollution).

Oxygen demanding wastes

Dissolved oxygen DO has been a fundamental requirement of life for the plant and animal population in any given body of water. Their survival is dependent upon the ability of the water to maintain certain minimal concentrations of this vital substance.

A body of water has been classified as polluted when the DO concentration drops below the level necessary for sustaining a normal biota for that water. The primary cause of water deoxygenation has been the presence of substances collectively called oxygen-demanding wastes. These are substances easily broken down or decayed by bacterial activity in the presence of oxygen. The available dissolved oxygen has been consumed by bacterial activity, and thus, the presence of such materials quickly gives rise to depletion of dissolved oxygen.

Oxygen-demanding wastes are primarily organic materials that are oxidized by bacteria to carbon dioxide and water. These substances have been deleterion because their decomposition gives rise to oxygen depletion in both freshwater and marine systems. They also produce anaesthetic odors, endanger water supplies, and decrease the recreational value of waterways.

Sewage and other oxygen demanding wastes have been classified as water pollutants because their degradation leads to oxygen depletion, which affects (and even kills) fish and other aquatic life; because they cause annoying odors; because they impair domestic and livestock water supplies by affecting taste, odors, and colors and because they may give rise to scum and solids that make water unfit for recreational use.

Although some inorganic substances occur in this category, most oxygen demanding wastes are organic compounds. Pollutants in this category typically come from such sources as sewage, both domestic and animal; industrial wastes from food processing plants; wastes from paper mill activities; tanning operation by products;

and effluent from slaughter houses and meat packing plants.

Most compounds involved in this type of pollution have carbon as their most abundant element. One reaction they undergo, with bacterial help, is the oxidation of carbon to CO_2.

$$C + O_2 \rightarrow CO_2$$

As oxygen-demanding wastes rapidly deplete the DO of water, it has been important to be able to estimate the amount of these pollutants in a given body of water. The biochemical oxygen demand (BOD) of water has been a quantity related to the amount of wastes present. In a water sample, the BOD indicates the amount of dissolved oxygen used up during the oxidation of oxygen demanding wastes. It could be found out by incubating a sample of water for five days at 20°C. The amount of oxygen consumed (BOD) could be found out by chemical determination of the DO concentration of the water before and after incubation.

A BOD of 1ppm has been characteristic of nearly pure water. Water has been regarded fairly pure with a BOD of 3 ppm, and of doubtful purity when the BOD value reaches 4 ppm.

Table 1
Comparison of Decomposition and Products Under Differing Conditions

Aerobic conditions	*Anaerobic conditions*
$C \rightarrow CO_2$	$C \rightarrow CH_4$
$N \rightarrow NH_3 + HNO_3$	$N \rightarrow NH$ + amines
$S \rightarrow H_2SO_4$	$S \rightarrow H_2S$
$P \rightarrow H_3PO_4$	$P \rightarrow PH_3$ and phosphorus compounds

The disappearance of plant and animal life is an obvious result of the oxygen depletion of water. This occurs by a direct killing effect or because of migrations to other areas.

The products of decomposition following these different pathways have been quite different, as shown in Table 1.

Methane (CH_4) has been odorless and flammable; amines are having a fishy smell; hydrogen sulfide is bad smelling and toxic; and some phosphorus compounds have unpleasant odors. When these contributions are added to the odor of decaying fish or algae, it becomes apparent that a shift from aerobic to anaerobic conditions of decomposition is not which is favored by user of fresh air.

Disease-causing agents

Water has been a potential carrier of pathogenic microorganisms and can endanger health and life. The pathogens most frequently transmitted through water have been those responsible for infections of the intestinal tract (typhoid and para-typhoid fevers, dysentery and cholera) and those responsible for polio and infections hepatitis. Historically, the prevention of water-borne diseases was the primary reason for pollution control in water. Modern disinfection techniques have greatly reduced this danger. This is not true for some large parts of the world, where for example, cholera epidemics are still common. The fact that such disease causing agents are under control must not result in a sense of false security. The occurrence of a polluted water supply leading to an outbreak disease is always a possibility. The responsible organisms are present in the feces or urine infected people and are ultimately discharged into a water supply.

Waste water released from municipalities, sanitoria, tanning and slaughtering plants, and boats may be sources of bacteria or other microorganisms which are capable of producing disease in men and animals, including livestock. Any fair-sized community at any given time has been likely to have some persons who are diseased so that disease microorganisms are almost always present in sewage.

There have been several types of human infections, not all of which are transmissible through water. Many of the diseases whose epidemics recurrently decimate human populations get transmitted by water, however, cholera and typhoid being important examples.

Table 2

Examples of Human Infections

A. Animal infections that are of public health importance because they are transmissible to man.

1. Tetanus from horses and cattle transmitted by inoculation or contract with animal feces.
2. Bubonic plague from wild rodents by insect (flea) bite.
3. Anthrax from herbivorous animals by direct contact.
4. Radies from dogs, bats, etc., by bites.
5. Bovine tuberculosis from cattle through ingestion or air-brone transmission.
6. Jungle yellow fever from monkeys through mosquito bites.
7. Several types of encephalitis from birds and fowl through mosquito bites.
8. Trichinosis from swine through ingestion.

B Primarily human infections in which the infective agent has a certain period of extrahuman residence before transmission.

1. Schistosomiasis ("snail fever") from water from snails.

2. Urban yellow fever from mosquitoes.
3. Hookworm from soil by skin penetration.
4. Malaria (also a mosquito infection) from mosquitoes.
5. Typhus from lice.

C. Infections that persists or multiply in the external environment and are transmissible from man to man.

1. Cholera, typhoid fever, bacillary dysentery, poliomyelitis, and infectious hepatitis from water and food through ingestion.
2. Staphylococcal and streptococcal diseases from food, air, and the proximate environment through contact and inhalation.
3. Smallpox from air, dust, and the proximate environment through inhalation.
4. Coxackie and ECHO virus diseases from water through ingestion.

The identification of pathogens in water needs very large samples and many sophisticated techniques and is too time-consuming and expensive for routine pollution tests. The standard method involves determination of the most probable number (MPN) of coliform organisms in the water sample.

Plant nutrients

Nutrients are an important limiting factor in the growth of all plants. With all other factors equal, the rate and profuseness of plant growth are proportional to the amount of nutrient available. Plant nutrients (phosphorus and nitrogen) enter fresh and marine system and lead to or intensify eutrophication of these systems. These nutrients tend to accumulate in groundwater since it is out of the euphotic zone. As the groundwater moves laterally and reaches the surface waters, these materials add to the nutrient levels already present.

Plant nutrients like nitrogen and phosphorus are able to stimulate the growth of aquatic plants, which get interfere with water uses and later decay to produce disagreeable odors and add to the BOD of the water. Excess algal growths all of particular concern because they cause low dissolved oxygen levels and create treatment problems for municipalities and industries, in addition to interfering with recreational uses. The enrichment of water with nutrients is a naturally occurring biological process called eutrophication. The term comes from two Greek words meaning "well nourished". This enrichment leads to other slow processes collectively referred to as natural aging of lakes. In some pollution articles, the term eutrophication includes both nutrient enrichment and lake-aging processes. The existence of peat and muck soils and deposits of coal and oil are proof that eutrophication and aging have been taken place in the past. The steps in eutrophication and again of a lake have been as follows:

1. Streams from a drainage basin gradually bring soil and nutrients to a newly formed lake, increasing the fertility of the lake water.
2. The increased fertility leads to an accumulating growth of aquatic organisms, both plant and animal.
3. As living matter increasing and organic deposits pile up on the bottom of the lake, it becomes more shallow, warmer, and richer in nutrients.
4. Plants take root at the bottom and gradually occupy more and more of the space. Their remains accelerated the filling of the basin.
5. The lake gradually becomes a marsh and finally a field or forest as it has been overrun by vegetation.

The time needed for this aging process to be completed could be measured in thousands of years. The actual time depends upon the size and mineral content of the basin, and the climate.

The growth of green plants, including algae, needs the availability from fifteen to twenty elements which, together with their chemical symbols, have been given in Table 3. The relative amounts of these elements needed for plant growth depend upon the species involved. The amounts needed for the growth of three typical nuisance algae are given in Table 3. The values have been based on chemical analyses of the dried algae.

Table 3

Elements Necessary for Plant Growth

Carbon (C)	Magnesium (Mg)	Boron (B)
Hydrogen (H)	Calcium (Ca)	Vanadium (V)
Oxygen (O)	Sodium (Na)	Chlorine (Cl)
Nitrogen (N)	Iron (Fe)	Molybdenum (Mo)
Phosphorus (P)	Manganese (Mn)	Cobalt (Co)
Sulfur (S)	Copper (Cu)	Silicon (Si)
Potassium(K)	Zinc (Zn)	

Table 4

Chemical Composition of Three Nuisance Algae

		Algae	
Element	*Microcystis*	*Anabena*	*Cladophora*
C	46.5wt%	40.7wt%	35.3wt%
N	8.1	9.4	2.3
P	0.7	0.71	0.50
K	0.8	1.2	6.1
Ca	0.53	0.36	1.7
S	0.27	0.53	1.6
Fe	0.27	0.08	0.23
Mg	0.17	0.42	0.23
Na	0.04	0.18	0.18
Mn	0.03	0.008	0.10

Zn	0.005	0.000	0.001
Cu	0.004	0.007	0.019
B	0.0004		0.0085

Most elements essential for plant growth have been available to the plant in amounts well in excess of the plant's needs. A few, however, have been present in amounts very close to the quantities required for plant growth. These may be used by the plant almost to the point of exhaustion.

Considerable interest gets centered around these elements because they appear to behave as natural controls in preventing excessive plant growth. The growth of a plant will cease when the least available element gets depleted.

Man's role in eutrophication appears to be one, of increasing the amounts of these growth-limiting elements. Great changes in the availability of some elements in water have taken place through man's activities.

Both nitrogen and phosphorus occur in small amount in natural waters, but their concentration get greatly increased by the activities of man. As much as 80% of the nitrogen and 75% of the phosphorus added to surface waters originate from man-made sources. Table 5 gives estimates of possible nutrient discharge (as P and N) into estuaries from a hypothetical community. The community has been made up of 10 million people and the nutrients are put into a river discharging 15,000 cubic feet of water per second into the estuaries.

It is regarded cubic feet of water per second into the estuaries. It is regarded that 20 volumes of seawater dilute each volume of discharged river water.

Table 5

Example of Possible Nutrient Discharges to Estuaries from Various Sources (Metric Tons Per Year)

	Nitrogen (N)	*Phosphorus (P)*
Sewage : population of 10 million, per capita generation 400 litres/day	6,000	15,000
River water : discharge 500 m^3/sec (approx. 15,000 cfs)	300	7
Subsurface seawater : 20 volumes of seawater mixing with each volume of river water	6,000	900
Storm-water runoff : 75cm (30 inches) per year area	60,000	800

The dominant source of phosphorus has been seen to be sewage, whereas nitrogen in various forms has been supplied in nearly equal amounts by sewage treatment plants, storm-water run-off, and subsurface seawater.

The phosphorus content of domestic sewage is becoming a topic of concern. It is estimated that up to 70% of this comes from the use of household detergents, and it should be subject to rather easy control procedures. The elimination of phosphorus compounds from detergents has been discussed at great length at another place.

Algae growth needs many different nutrients like carbon dioxide, nitrogen, phosphorus, iron, manganese, boron, cobalt, vitamins, hormones, etc. The elimination of any one of the essential nutrients would not allow algae growth but there is an active scientific controversy about which nutrients should be controlled. It is generally felt that control of nitrogen and phosphorus would be best but some scientists think that the availability of carbon is a much more important limitation, and the large amounts of phosphorus present in the sediments in lakes and rivers yield a vast reservoir available for the growth of algae.

Organic chemicals

The exotic organic chemicals include surfactants in detergents, pesticides, various industrial products, and the decomposition products of other organic compounds. Analysis of polluted waters reveal the presence of a wide variety of these compounds. Some of these compounds have been found to be toxic to fish at very low concentrations, such as 1 ppm phenol. Many area not biologically degradable, or are degraded only very slowly.

Laundry detergents, a common constituent of waste water, have been another organic contaminant. Detergents consist of two major components : a surfactant or sudsing agent and a series of builders. The surfactant or sudsing agent can become a problem as it is not readily broken down by bacterial action and has a long residence time. Even in relatively low concentrations, it can cause water to form foam.

Most detergents manufactured in the early 19.0s were composed of approximately 40% sodium tripolyphosphate, the "builder" component. It is calculated that such phosphorus-bearing builders account for one half the total phosphate in most wastewaters. Due to this extremely high phsophorus concentration, these detergents are a major source of eutrophication in freshwater systems. In addition as detergents have been contributing a large amount of phosphorus to wastewater without a corresponding addition of nitrogen/phosphorus imbalance will take place when this material enters either marine or fresh-water systems.

Hydrocarbons, in the form of gasoline and motor oil, although insoluble in water, have been carried from roadways and parking areas in a rainwater runoff.

In addition to the substances mentioned above, analysis of water reveals the presence of a wide variety of other organic chemicals. Some of these compounds are known to be toxic at very low concentrations, while many are not biodegradable or break down very slowly.

Oil

The production, distribution, and use of such large yearly quantities results in some oil contamination of the environment. Some of this contamination is accidental, some is not.

Inorganic chemicals and mineral substances

The category of water pollutants have been including inorganic salts, mineral acids, and finely divided metals or metal compounds. These substances enter natural waters because of activities in various smelting, metallurgical, and chemical industries; mine drainage and various natural processes. The presence of these brings about three general effects: the acidity, salinity, and toxicity of the water may be increased.

Inorganic chemicals of many types enter water from municipal and industrial waste waters and urban runoff. These pollutant are able to kill or injure fish and other aquatic life and they can interfere with the suitability or water for drinking or industrial use. Many of these inorganics are not only toxic but tend to concentrate in food chains.

A prominent example has been the occurrence of mercury in water. A number of industrial processes make use of mercury, some of which is eventually disposed of in waste-water effluents. It is now known, that anaerobic bacteria in bottom muds can convert inorganic mercury into methyl mercury (CH_3Hg^+), which can be concentrated in living things and lead to mercury poisoning.

One potential pollutant arises in petroleum drilling, which gets discharged along with crude oil when the latter is pumped to the surface. In some-places the brines have proved valuable sources of important minerals and elements, such as bromine, iodine and magnesium.

Another very important problem, is acid mine drainage. On exposed coal mine surfaces, minerals, containing sulphur (most notably iron pyrite, FeS_2) come into contact with air and water, forming sulfuric acid that has been carried into stream by waters draining from the mines. This takes place from abandoned mines as well as operating mines and has been most pronounced in bituminous coal mines.

Acid mine drainage has been the primary source of pollutants that increase water acidity. This drainage adversely affects thousands of miles of streams and has been one of the most significant causes of water quality degradation in coal-producing areas. The actual pollutants present in mine drainage have been sulfuric acid (H_2SO_4) and soluble compounds of iron. These substances are formed because of a reason between air, water, and pyrite (FeS_2) present in coal seams. Certain types of bacteria have been involved in the reaction, but their role has been not completely understood. This reaction can occur in both underground and surface mines.

During mining operations, in deep mines the strata between the coal seam and the surface have been invariably disturbed. Fissures appear through which water drains into the mine from many surface areas. This water containing the harmful pollutants is eventually discharged into surface streams, either naturally or through man-made processes. Damaging mine drainage is formed in surface mines if the surface run-off water comes in contact with pyrite-containing coal.

About 60% of the mine drainage pollution problems originate in mines once worked but now abandoned.

Large amounts of acid have been used to clean oxides and grease from metals. The used acids from these pickling operations have been becoming serious problems in some localized areas.

Due to the presence in solution of carbonate (CO_3^{2-}) and bicarbonate (HCO_3^-) ions, the pH of most productive, fresh natural waters lies between 6.5 and 8.5. This natural buffer system makes it possible for small amounts of acid or base to dissolve in the waters without making appreciable changes in the pH. These ions also represent an indispensible reservoir for carbon needed by aquatic plants during photosynthesis. Such use of these ions by plants can influence the buffering capacity of water, as there is a decided limit on the rate at which CO_2 (the source of CO_3^{2-} and HCO_3^-) can be obtained from the atmosphere to replace that used by plants.

Large influxes of strong acid have been able to overwhelm this buffering ability of water and bring about drastic drops in pH values. The effects of these changes depend upon the magnitude of the pH drop involved. Some effects have been as follows:

(a) *Aquatic life destruction*. At pH levels below 4.0, all vertebrates, a most invertebrates, and many microorganisms get destroyed. Most higher plants are eliminated, leaving only a few algae and bacteria. Acid mine drainage is one of the primary cause of fish kills. Excessive precipitation increases the mine drainage output and compounds the problem. During high water seasons (winter and spring), the pH of the water in some areas has been reported to be as low as 2.5.

(b) *Corrosion*. Water having a pH lower than 6.0 can bring about excessive corrosion of plumbing systems, boats, piers, and related structures.

(c) *Agricultural crop damage*. Acidity and alkalinity of irrigation water have been usually of little consequence over a pH range of 4.5-9.0, as the soil has been a buffered system. Problems may arise if the pH drops below 4.5. Such acidic water increases the solubility of such substances as Fe, Al and Mg salts. These ions, at the resulting high concentrations, have been sometimes toxic to plants.

Large amounts of salinity in water cause problems other ''han those related to human consumption. Dissolved inorganic and mineral substances exert abverse effects on aquatic animal and plant life and cause many irrigation problems in the agricultural industry. Damage to aquatic life is primarily related to the osmosis process, assuming the dissolved substances are nontoxic. Generally, the concentration of dissolved materials in body fluids is the maximum that an aquatic organism can tolerate. When these organisms are in contact with water containing higher concentrations, there is a tendency for water to move out of the cells of the organism into the surrounding water. The resulting increase in concentration within the cells of the organism can lead to death. Many fresh water species disappear when waters become brackish.

One of the most serious long term effects of increased salinity of water involves the use and re-use of water in irrigation. It has been estimated that about 25% of the irrigated land of the world is now affected to some degree by water salinity. Irrigation water brought onto a field always contains some dissolved salts. The concentration of these salts is in the range of 25 8000mg/l. Plants extract water from the irrigated field, but most of the dissolved salts are excluded by the roots. Water that evaporates from the soil surface leaves dissolved salts behind. These two processes make residual salts to accumulate in the soil.

The toxic properties of numerous inorganic compounds, especially those of some of the heavier metallic elements, have been known for years. Some of these compounds are having desirable properties and so are routinely manufactured. Appropriate precautions are taken to insure the safety of individuals involved in the manufacturing process. The use of these compounds give rise to their introduction into the environment either directly or indirectly, intentionally or unintentionally. The recent detection of these metallic substances in air and water in concentrations approaching toxic levels has created a great amount of concern. The most toxic, persistent, and abundant of these compounds in the environment appear to be those of the metals mercury (Hg), lead (Pb), arsenic (As), cadmium (Cd), chromium (Cr), and nickel (Ni).These metals accumulate in the bodies of organisms, remain for long periods of time, and behave as cumulative poisons. More research is needed to determine the seriousness of these potential hazards.

Sediments

Sediments are soil and mineral particles which are washed from land by storms and floodwaters, from croplands, unprotected forest soils, overgrazed pastures, strip mines, roads, and bullozed urban areas. Sediments are able to fill stream channels and reservoirs : erode power turbines and pumping equipment; reduce the amount of sunlight available to green aquatic plants; plug water filters; and blanket fish nests, spawn, and food supplies, thus reducing the fish and shelfish populations.

Sediments are almost regarded as a type of pollution due to the naturally occurring process of erosion. Sediments produced by that process do represent the most extensive pollutants of surface waters. It is estimated that suspended solid loadings reaching natural waters have been at least 700 times as large as the solid loadings from sewage discharge.

Sediment production is a tedious problem in some parts to the world. This variation is not all caused by man's activities, but also involves the type of soils, geology, topography, precipitation, and vegetation cover. The detrimental effects of sediment in water are:

(a) *Stream channels, harbors, and reservoirs are filled.* This makes channels to overflow more easily, changes flow rates and depths of channels, and reduces the useful life or reservoirs. Expensive dredging is needed to counteract these effects.

(b) *Destroys aquatic animals.* Sediment setting on the bottom lowers fish and shelfish populations by blanketing fish nests and food supplies.

(c) *Reduces light penetration into water.* The reduction in the amount of sunlight penetrating into the water lowers the rate of photosynthesis by plants, which in turn results in a decreased production of oxygen needed for normal stream balance.

(d) *Water is clouded.* It increases the cost of treating water used for culinary purposes. Sediment passing through power plant turbines makes serious abrasion and wear. The hunting ability of fish get curtailed if they depend on their vision.

Radioactive materials

Harmful radiation may result in water environments

from the wastes of uranium and thorium mining and refining, from nuclear power plants; and from industrial, medical, and scientific utilization of radioactive material.

Many radioactive substances are lethal at relatively low concentrations and in minute a mounts may be mutagenic.

Perhaps the greatest problem of radioactive pollution, as a result of uranium production, has been caused by the large quantities of "uranium tailings" produced. This finely divided solid material remains after useful materials have been leached out. Huge piles of these tailings are found in uranium producing areas. These tailings create a radiation pollution problem as they contain radioactive decay products of uranium. Two of these radioactive materials are thorium-230 ($^{230}_{90}Th$) and radium-226 ($^{226}_{88}Ra$). Substances like these can be dissolved or eroded from piles of tailings by rainfall, and thus they can get mixed up into the general water supply. Radium and thorium is chemically similar to calcium and so get absorbed by the bones when taken in the body. Some waters in the Colorado River Basin have had the concentration of $^{226}_{88}Ra$ increase to about double the maximum level permissible for human consumption.

An important source of radioisotopes is nuclear weapons testing. The amount and variety of radioactive materials formed depends on the type of weapon tested. Some of the radio-isotopes have a very short half life and last only a few seconds or minutes, and some others may have a half-life of several hundred years. These materials reach the earth as radioactive fallout. Before the first test explosion were conduced, it was thought that fallout (dust and debris) would fall to earth quickly and at no great distance from the detonation site, thus the spread of radioactivity would be limited. However these assumptions have proved wrong, especially in the case of large explosions.

Radioactive particles can, remain suspended in the air long enough to circle the globe many times, dispersing as they go. It is true however, that the main effects of atmospheric testing are found in the immediate vicinity of the detonation site. Because of concern about such fallout in 1963, of the Limited Nuclear Test Ban Treaty was signed. According to the U.S. and U.S.S.R. will limit testing of nuclear weapons to underground detonations. The purpose of underground detonations, carried out in the U.S. at the Navada test site, is to limit the amount of atmospheric fallout, and in majority of cases this is achieved. Occasionally accidental releases of radiation reach the atmosphere due to "venting" at the explosion site. When this happens, radioactive gases are forced up through the hole used to place the explosives. The leakage may take place through channels left the control and monitoring cables, natural cracks or fissures in the rock, fractures formed by the blast, or combinations of these. Sometimes venting occurs when the materials used to fill the hole are simply blown out. Leakage may immediately take place after the explosion or later, when rock and earth fall into the cavity formed by the explosion and leave a channel for gas venting.

In only 12 of 190 tests conducted at the Nevada site from 1961 through 1969 ventings were reported different amount of radiation are released. It is released near ground level and so poses no world-wide fallout problems, but it may result in very serious local hazards.

Strontium-90 ($^{30}_{38}Sr$), a component of radioactive fallout which has a half-life of 28 years, is chemically similar to calcium. Calcium is absorbed from the soil by plants and passed on to animals, where it is used in the formation of bones and teeth. Man gets calcium from both plant and animal sources like milk, vegetables, and cereal grains. Strontium is produced from these sources and, due to chemical similarities to calcium it is also deposited in bones and teeth. The marrow of bones is the main site of blood-cell formation. The presence of radioactive strontium-90 in the surrounding bone tissue seriously curtails this which may result in production anemia or more serious disorders.

Cesium-137 (Cs) which is chemically similar to potassium is a common constituent of all living cells. Cesium from fallout passes to man through contaminated meat and dairy products or contaminated grains and leafy vegetables. Cesium contamination of foods results in the same effects as strontium contamination. The soft parts of the body, specially the muscles, are badly affected by the presence of cesium.

More and more of nuclear energy is being produced so as to meet the ever increasing demand for electrical power. Nuclear generators are better as compared to more commonly used types. Because, nuclear energy is cheaper to use than that from conventional sources such as fossil fuels or falling water. Moreover, nuclear generators do not produce any SO_2 or particulate pollutants commonly produced by fossil-fuel burning generators. By the year 2000, nuclear power is expected to be the main source of electric power generation.

Nuclear power plants are different from conventional electricity generating facilities. In the conventionally fueled power plant, fossil fuel is burnt to produce heat while in a nuclear plant the heat is generated by a nuclear reaction. In both the heat produces steam which turns a generator and produces electricity.

Four types of pollutants that are formed by nuclear generating plants are as under:

(a) *Low level radioactive liquid wastes.* Radioactive isotopes are produced when impurities in the primary coolant water and corrosion products from coolant pipes are bombarded with neutrons from the core area. This can be avoided to some degree by using demineralized coolant water.

(b) *Heat.* The secondary cooling system takes away huge amounts of heat from the reactors and mixes it into natural water supplies.

A variety of research laboratories, including medical and biochemical facilities, release radioactive wastes into the air and sewer systems. It is not now feasible to collect and dispose of these materials. Nor is it feasible to collect and bury the large amounts of radioisotope-containing liquids, solids, and gases forward by laboratory uses other than research, like diagnostics or treatment. Presently, the amounts of radioactive waste discharged from such laboratory sources poses no threat to natural communities of plants, animals, and man.

(c) *Liquid and gaseous wastes from fuel elements.* Complete sealing of the fuel in steel or zirconium containers is apparently impossible to attain or sustain Minute cracks allow fission products to escape into the primary coolant. This further complicates the disposal problem of low level wastes discussed earlier.

(d) *Fission products.* Within 1-3 years, fission products (the ashes of nuclear fuels) accumulate to the point that they absorb sufficient neutrons to slow or stop the chain reaction. Then the extremely radioactive fuel elements are removed and shipped in special containers to a fuel reprocessing plant. In these plants the fission products are separated from the remaining usable fuel. The fuel is returned to the reactor where it is used and the waste fission products are stored, in liquid form, in huge underground stainless steel tanks. Hundreds of millions of gallons of these high level wastes are now in storage.

Heat (Thermal pollution)

Vast amounts of water find use for cooling purposes by steamelectric power plants (and other industries to a lesser extent). Cooling water is discharged at a raised temperature, and some rivers may have their temperatures so high (even up to 40°) that fish life completely gets eliminated and the river becomes useless for assimilation or pollution further cooling. Increasing the water temperature of a system is harmful since it generally alters the chemical, physical and biological characteristics of that system. In addition to the possibility of decreasing or eliminating various aquatic forms, it may also stimulate spawning at a time of year when food supplies are limited. This leads to starvation of the newly spawned individuals in the population. High temperatures also decrease the density and viscosity of water, causing an increased settling rate of suspended solids. Evaporation rate is increased and, in marine systems, this may lead to localized areas of abnormally high salinity.

Heat is not ordinarily thought of as a pollutant by many people, at least not in the same sense as a corrosive chemical. However, the addition of excess heat to a body of water brings about adverse effects as numerous as many of the chemical pollutants. This serious problem of thermal pollution originates primarily with the practice of using water as a coolant in many industrial processes. Most water used for this purpose gets returned, with the added heat, to the original sources. At present, about 70% of the water diverted, to industrial use serves as a cooling medium.

Used coolant water frequently may be having a temperature 20°F higher than the river or stream to which it gets returned. This is added heat raises the temperature of the natural waters, with the results that : (1) the amount of dissolved oxygen in the water gets decreased; (2) the rates of chemical reactions have been increased; (3) false temperature cues have been given to aquatic life; and (4) lethal temperature limits may be exceeded.

The addition of heated water to a cooler, body of water may accelerate the lowering of DO levels due to density differences between the two. The less dense warm water tend to form a layer on top of the cooler, more dense water. this occurs particularly when the body of cool receiving water is deep. The resulting blanket of "hot" water cannot dissolve as much atmospheric oxygen as the underlying cold water, which is denied contact with the atmosphere. Normal biological reduction of the DO level of the atmospherically unreplenished lower layer may give rise to anaerobic conditions.

Another effect of this stratification may show up down-stream from a dam when the oxygen deficient lower level get discharged through the lower gates of a dam. Serious effects on downstream fish life may result. Also, the ability of the stream below the dam to assimilate oxygen demanding wastes will get curtailed.

In thermally polluted water, fish need more oxygen due to an increased respiration rate. However, the available oxygen in such water gets decreased. Thus, thermal pollution affects fish in a double-barreled fishion.

Other reactions are also influenced. Trout eggs hatch in 165 days when incubated at 37°F. When water temperatures have been 54°F, only 32 days are needed and no hatching takes place at water temperatures in excess of 59°F. Such a result can be disastrous to fish populations.

The life cycle and natural processes of many aquatic organisms have been closely and delicately geared to water temperature. Fish often migrate, spawn, and are otherwise distributed in response to water temperature cues. Shellfish, such as oysters, spawn within a few hours after their environment reaches a critical temperature. These normal life patterns of aquatic organisms can get completely disrupted by artificial changes in water temperatures. Fish have been found to vary widely in their water temperature preferences. An indication of this has been the commonly used classification of fish as either cold- or warm-water species. An upper lethal, temperature limit exists for each species. It is known that temperatures well below the lethal value can bring about stress in organisms; hence the much lower recommended values. Lethal limits can be exceeded, in some examples under conditions of warm water waste discharge.

Another factor to be considered in addition to the temperature change has been the rapidity with which the change occurs. Fish are able to acclimate themselves to moderate temperature changes (below lethal levels) if the change has been not a sudden one. Normally a moderate temperature change over a 30-40 hour period could be handled by the fish. For instance, 95% of the eggs of largemouth bass perish when suddenly transferred from water at 65-70°F into water at 85°F. However, if similar eggs are kept in water and the temperature gradually increased to 85°F over a 30-40 hour time period, 80% of the eggs survive.

An additional threat to aquatic life has been created by the common practice of chlorinating coolant water prior to use. This is carried out to prevent bacterial growths that clog pipes. The chlorine affects organisms in the area receiving the used water until dilution occurs. The chlorine has been able to kill microorganisms important in some food chains and may in this way exert a negative influence on fish populations. A problem of thermal population not related to aquatic life has been the reduced cooling capability of warmed waters. It is important because nuclear reactor power sources need about 50% more cooling water for a given temperature increase than similar power plants using fossil fuels, and it appears that nuclear plants will find use extensively in the future.

Different solutions to the thermal pollution problem has been postulated and some have been already in use. Several industries have incorporated cooling towers into their operations to remove heat from cooling water before returning it to the natural water supply. Two types of towers, wet and dry, have been used. Water is run over baffles in a thin layer in the wet tower. Cool air, entering at the bottom, circulates upward and removes heat from the water. In dry towers, air is forced over water containing pipes by huge fans. Heat gets exchanged by radiation and convection from the pipes.

The use of cooling ponds or lakes has been another alternative. These ponds could act as ice-free wintering area for waterfowl in northern locations or as a means of extending the range (northward) of certain fish. Both of these have been positive results of thermal pollution.

Sources of contamination of water pollution

Introduction

The major sources of water contamination have been domestic, industrial and agricultural waste, as well as solid waste thermal pollution, shipping water pollution and radioactive waste.

Domestic water pollution

It includes waste water from homes and commercial establishments. Domestic waste water arises from many small sources spread over a fairly wide area but is transmitted by sewers to a municipal waste treatment plant.

Generally, the impurities in domestic wastes get diluted and seldom total more than 0.1% of the total mass. This material has been largely organic and gets oxidized by bacterial decomposition to nitrate, phosphate, carbon dioxide, and water. As this type of decomposition needs the use of dissolved oxygen, it place an oxygen demand on the system. Because of this tendency to remove oxygen in the decomposition process, a common indicator used to monitor this type of input in receiving water has been the BOD (Biological Oxygen Demand) Test. In this analysis the amount of oxygen for decomposition is measured over a 5-day period. In systems receiving significant amount of organic material, the bacterial decomposition will remove large amounts of dissolved oxygen, which causes oxygen depletion. These systems are considered to have a high BOD. Conversely, when the input of contaminants is absent or minimal, the oxygen levels will not be drastically reduced because the bacteria will not need an excessive amount of oxygen to decompose small amounts of materials. In these cases there would be a low BOD on the system.

Industrial water pollution

This occurs in large amounts in specific locations, making collection and treatment fairly simple to accomplish. There are water-using factories which are discharging wastes with a total BOD load about three to four times as large as the load from the sewered population. Only about 7 or 8% of industrial waste waters have been disposed of in municipal sewer system.

Wastes from textile manufacturing processes could be generated from the washing out of the impurities in the fibers, as well as in the discarding of chemicals used in the processing of the fibers. Generally, these wastes have been organic, have a high BOD, and are extremely alkaline.

Food-processing wastes from meat, dairy, and sugar-beet processing, as well as brewing, distilling and canning orations, generate large amounts of organic by-products that have been disposed of in waste-water. When the waste-water is discarded, along with these by-products, it leads to high BODs and a consequent oxygen depletion in the receiving water via the same bacterial process involved in the decomposition of domestic wastes.

The effluent released from pulp and paper processing operations has been a mixture of chemicals used in the digestion of raw wood chips, cellulose fibres, and dissolved lignin. This waste-water also has paper and wood preservatives, like pentachlorophenol and sodium pentachlorophenate, as well as methyl mercaptan, all of which have been toxic to fish. This effluent has been brownish in colour and lowers the photosynthetic rate of aquatic communities by hindering sunlight penetration into the water column. Consequently, the organic wastes from these plants increase the BOD of the receiving water, while colour imparted to the water interferes with sunlight penetration' reduces photosynthesis, and further lowers oxygen levels.

Metal industries keep a wide array of contaminants in their waste-water. The specific contaminants and concentrations depend solely on the particular manufacturing process employed. For example, steel mills employ and contaminate water in the coking of coal, the pickling of steel, and the washing of flue gases from blast furnaces. These waters, after use tend to be acidic and have various deleterious substances, like phenol, cyanogen, ore, coke, and fine suspended solids. Other industries release traces of the metals produced or plated in their waste-water. Metals commonly found in these waste-waters have been chromium, mercury, nickel, lead, copper, and cadmium.

A variety of contaminants enter marine and fresh water environments in the effluents release from the various chemical manufacturing plants. The release of acids results not only from acid manufacturing processes but from practically all other chemical manufacturing processes as well. In addition, synthetic fibres (e.g., rayons), bases, pesticides, and other organic and inorganic chemicals have been added, depending on the products being manufactured.

Most of the Indian rivers and fresh-water streams are seriously polluted by industrial wastes or effluents which come along waste-waters of different industries such as petrochemical complexes; fertilizer factories; oil refineries; pulp, paper, textile, sugar and steel mills, tanneries' distilleries, coal washeries, synthetic material plants for drugs, fibres, rubber, plastics, etc. The industrial wastes of these industries and mills include metals (copper, zinc, lead, mercury, etc.)., detergents, petroleum, acids, alkalies, phenols, carbamates, alcohols, cyanide, arsenic, chlorine and many other inorganic and organic toxicants. All of these chemicals of industrial wastes have been toxic to animals and may bring about death or sublethal pathology of the liver, kidneys, reproductive systems, respiratory systems, or nervous systems in both in vertebrate and vertebrate aquatic animals (Wilbur, 1969). Chlorine which is added to water to control growth of algae and bacteria in the cooling system of power station, may persist in streams to cause mortality of plankton and fish. Heavy fish mortality in river Sone near Dehrion-on-sone in Bihar is reported to cause by free chlorine content of the chemical wastes discharged by factories near Mirzapur in U.P.

Table 6
Some Indian Rivers and their Major Sources of Pollution

Name of the River	*Sources of Pollution*
1. Kali at Meerut (U.P.)	Sugar mills; distelleries; paints soap, rayon, silk, yarn, tine and glycerine industries.
2. Jamuna near Delhi	D.D.T. factory, sewage, Indraprastha Power Station, Delhi.
3. Ganga at Kanpur	Jute, chemical, metal and surgical industries; tanneries, textile mills and great bulk of domestic sewage of highly organic nature.
4. Gomti near Lucknow (U.P.)	Paper and pulp mills; sewage.
5. Dajora in Bareilly(U.P.)	Synthetic rubber factories.
6. Damodar between Bokaro and Panchet	Fertilizers, fly ash from steel mills, suspended coal particles from washeries, and thermal power station.
7. Hoogly near Calcutta	Power station; paper pulp, jute textiles, chemical mills, paint, varnishes, metal, steel, hydrogenated vegetable oils, rayon, and soap, match, shellac, and polythene industries and sewage
8. Sone at Dalmianager (Bihar)	Cement, pulp and paper mills,
9. Bhadra (Karnataka)	Pulp, paper and steel industries.
10. Coccum, Adyar and (Bukinghum canal Madras)	Domestic sewage, automobile workshops.

11.	Cauvery (Tamil Nadu)	Sewage, tanneries,distilleries, paper and rayon mills.
12.	Godavari	Paper mills.
13.	Siwan (Bihar)	Paper, sulphur, cement, sugar mills.
14.	Kulu (between Bombay and Kalyan)	Chemical factories, rayon mills and tanneries.
15.	Suwao (in Balrampur)	Sugar industries.

Mercury like other heavy metals such as lead and cadmium has cropped up as a toxic agent of serious nature. Mercury, a by-product of the production of vinyl-chloride, is used in many chemical industries and it is also a by-product of some incinerators, power plants, laboratories and even hospitals. In Japan, illness and even death occurred in the 1950s among fishermen who ingested fish, crabs, and shell-fish contaminated with methyl mercury from Japanese coastal industries. This mercury poisoning produced a crippling and often fatal disease called Minamata disease. Initial symptoms of minamata disease included numbness of the limps, lips, and tongue, impairment of motor control, deafness, and blurring of vision.

Cellular degeneration occurred in the corebellum, mid-brain, and cerebral cortex and this led to spasticity, figidity, stupor and coma. In Japan in 1953, due to Minamata disease 17 persons died and 23 were become permanently disabled.

In India, all the 14 major rivers have become polluted. The river Damodar is perhaps the most heavily polluted river. River Mini-Mahi and Baroda has been other heavily polluted river which is having a variety of industrial and petrochemical wastes. The river Cooum flowing through Madras has been got polluted by sewage so much that not even the zooplanktons have been able to thrive in it. One litre of Cooum water is having a much as 900 mg or iron, 275 mg of lead, 1313 mg of nickel and 32 mg of zinc. Besides heavy metals, very high level of phosphates, slicates and nitrates also occur in the water. Sulphate levels ranging from 80-408 mgl^{-1} were the highest recorded among Indian rivers. The river Ganga from Hardwar to Calcutta is regarded as one unending sewer which is fit only to carry urban liquid waste, half burnt dead bodies, carrion, pesticides and insecticides. Nearly 312 industrial units are dumping their waste into the river, only a dozen have effluent treatment facilities. The 27 cities contribute 902 million litres of waste-water to the river each day. The water of Ganga affect the health of 250 million people of northern India.

Many of our lakes, notably the Dal lake, are becoming darkened, smelly and choked with excessive growth of algae.

Agricultural water pollution

It includes sediments, fertilizers, and farm animal wastes. These pollutants can all enter waterways as runoff from agricultural lands but farm animal wastes are an especially large problem near the large feedlots on which thousands of animals are concentrated.

Agricultural waste includes the pesticides that are sprayed on crops, as well as sediment, fertilizers, and plant and animal debris that are carried into waterways during periods of rainfall or as runoff and during the irrigation of farmland. Wastes generated by farm animals are also included in this category. The practice of keeping large numbers of animals in a small area has led to an excess of animal wastes generated in and confined to a given area, where it has been economically impossible to distribute wastes for reuse as fertilizers. These materials become a problem when they are allowed to enter waterways during the cleaning of the confinement areas or during periods of heavy rainfall, when runoff carries them into adjacent waterways. Since these wastes are organic, they increase the BOD of the receiving waters.

Inorganic fertilizers, being plant nutrients, lead to over-fertilization of waterways when they enter these systems through runoff or during irrigation. The addition of excess plant nutrients can lead to a disturbance of the phosphorus/nitrogen balance in these systems as well as excessive plant growth. When the plants die, they settle to the bottom and, since they are organic, increase the BOD of the system during decomposition. Soil erosion poses a fourfold problem. It increases the normal rate of filling of the waterways into which it washes, decreases the amount of fertile land for crop production, carries pesticide coated soil particles in the water, and decreases the transparency of the water, which limits photosynthesis. In addition, the sediment carried into fresh water systems tends to clog the gills of adult fish and settles out over incubating eggs, causing suffocation.

A remarkably large number of pesticides have come into widespread use in recent years. Many of these compounds have been not only non-biodegradable but are also only slightly soluble in water. Consequently, when sprayed on cropland they remain in the soil for long periods of time. During periods of heavy rainfall or when the crops are irrigated, they tend to be carried, as suspended particles, in surface, marine, or groundwater systems. In both fresh and marine systems they enter the food chain, undergo concentration in nontarget organisms, and increase in animal tissue to alarming levels. In surface, fresh, and groundwater systems they may also enter the drinking-water supplies of various communities.

Solid waste pollution

Solid waste varies in composition with the

socioeconomic status of the generating community. The following materials could be classified as solid waste:

(a) Garbage, which includes all decomposable wastes from households, as well as from food, canning, freezing, and meat-processing operations that are not disposed of in wastewater.

(b) Rubbish includes all nondecomposable wastes. These materials may be either combustible or noncombustible. Combustible materials would include garden wastes, cloth, and paper. Non-combustible materials include masonry, some chemicals, metals, and glass.

(c) Sewage sludge is generated from the settling processes in primary, secondary, and tertiary treatment methods as well as the solids from cesspools, which must be removed periodically.

(d) Miscellaneous materials include industrial wastes, like chemicals, paint, and explosives, as well as mining wastes, such as slag heaps and mine tailings.

The disposal of solid waste poses many problems, depending upon both the type of waste and the disposal method employed. The majority of the waste classified as combustible-rubbish garbage, and sewage sludge—has been disposed of by one of three major methods: incineration, using it as landfill, or disposal by ocean dumping. Incineration generally gives to air pollution; landfill operations or ocean dumping lead to water contamination. If landfill disposal is used, the material, as it decomposes, will dissolve in or become suspended in the rainwater percolating into the ground and thus into subsurface aquifers. This tends to contaminate not only groundwater but also surface waters, since the two systems are ultimately interconnected. Because of the highly organic nature of this material, a large BOD is placed on the receiving waters, and the sediment becomes coated with a highly organic ooze. Non-combustible materials are generally disposed of in landfill sites or by ocean dumping.

Thermal pollution

Thermal pollution takes place because many electric generating companies use water in the process of cooling their generators. This heated water is then released into the system from which it was drawn, causing a warming trend of the surface waters. Thermal pollution results when the heated effluent is released into poorly flushed systems. In these cases permanent temperature increases often result, which tend to decrease the solubility of dissolved oxygen. In lakes it also becomes possible to bring about nutrient redistributions and prolong summer stagnation periods.

When heated water gets released into large, well-fished marine systems there is little if any permanent temperature rise. There are, however, problems related to the operation of plants utilizing marine waters in the cooling process. Evidence reveals that seawater tends to corrode the cooling pipes, which are generally constructed of a copper nickel alloy termed Monel. These metal readily dissolve in the heated seawater and are then released into the marine environment together with the heated effluent. This adds to the nickel and copper concentrations of these systems. In addition, the screens covering the water-intake pipes rapidly foul with marine organisms, which decreases the flow of water into the plant. The screens have been commonly cleaned by using a concentrated detergent solution or copper sulfate. These cleaning materials have been then released into a contaminated the surrounding waters.

Shipping water pollution

It includes both human sewage and other wastes, the most important of which has been oil.

Oil pollution results from accidents involving oil tankers and from spills at offshore oil drilling sites. A more persistent source of oil pollution results from the practice of oil tankers, after they deliver the oil, to fill the empty tanks with seawater to act as ballast for the return trip. Prior to docking, the seawater ballast, contaminated with the oil that remains in the tanks, is discharged. Although this practice is illegal, it is difficult to prevent. The alternative is to pump the contaminated seawater into tanks at port. This is not only inconvenient but uneconomical. In addition, there will remain the problem of disposal when the disposal tanks are full.

Radioactive waste pollution

The major sources of radioactive wastes have been nuclear explosives, accidents at nuclear power plants, fuel-reprocessing plants, and research laboratories and hospitals that release these wastes into the atmosphere or into wastewater. Presently, most interest centers on radioactive iodine and strontium, since man is at the end of the food chains that concentrate these elements. The significance of the various parameters discussed and their use as water quality indicators has been summarized in Table 7.

Table 7

Water-Quality Indicators

Parameter 1	*Significance* 2	*Level* 3
Dissolved oxygen	General indicator of water quality; source of O_2 for respiration	Minimum acceptable level, 4-5mg/liter;10-15 mg/liter for reproduction of desirable fish

Total dissolved solids	Clog fish gills, bury eggs, reduce light penetration, increase heat absorption	Dependent on location
Total dissolved solids	Represents total mineral content which may or may not be toxic	A maximum of 400mg/ liter for diverse fish populations
BOD	Amount of dissolved oxygen removed during decomposition of organic matter in a given time; a general indicator of contamination due to biodegradable organics	BOD Water Status 1. mg/liter Very clean 2. mg/liter Clean 3. mg/liter Fairly clean 5. mg/liter Doubtful 10 mg/liter Contaminated
COD	Indicates the concentration of materials oxidizable by chemical reaction	0-5 mg/liter indicates very clean streams
pH	Indicates the addition of acids or bases	pH depends on actual system
Iron	Excessive amounts can clog fish gills;indicates drainage from iron-bearing sediments, mines, industrial processes	A maximum of 0.7 mg/ liter for diverse fish populations
Manganese	Concentration low in natural systems due to low solubility; high concentrations indicates contamination	A maximum of 1 mg/ liter is a common criterion for stream quality
Copper	Indicates drainage from copper-bearing sediment, mines, plating or other industrial sources	A maximum of 0.01-10 mg/liter is a common criterion for stream quality
Zinc	Indicates mine drainage or industrial input	A maximum of 1 mg/ liter is a common criterion for stream quality

Hg,Cd,Pb, Ni, Cr Ag, etc.	Indicates industrial input	A maximum of 1 mg/ liter is a common criterion for stream quality
Nitrate	A major plant nutrient; in high-concentrations it can promote exessive plant growth; major sources are fertilizers, sludge, and sewage.	A maximum of 0.3 mg/ liter to prevent excessive fertilization of streams
Phosphate	A major plant nutrient; major sources are detergents, fertilizer, sewage	A maximum of 0.03-0.40 mg/liter inorganic phosphate is a common criterion

Mercury

Mercury compounds, like many others, are dispersed throughout rocks, soil, air, water, and living organisms by a complex system of physicals chemical and biological processes.

Mercury has been used in a wide variety of ways. the largest user has been seen to be the chlor-alkali industry, which produces chlorine (Cl) and caustic soda (NaOH) by the electrolysis of salt (NaCl) solutions. The function of mercury in this process is to serve as the cathode of the electroysis cell.

The second greatest consumption of mercury has been in the production of electrical apparatus. This application encompasses a large variety of uses.

The use of mercury and its compounds as fungicides constitutes the third largest category of mercury consumption. In this application, use is made of the toxicity of the substances toward living organisms. These substances have been used to destroy fungi in the paint, pulp and paper, and agricultural industries. Mercury compounds have been added to paint that is to be used in humid regions to act as a latex preservative and a mildewcide. Paints for use on ships must contain "antifouling" agents to prevent barnacle growth. Mercury oxide (HgO) is often used in this application. Mildew proofing of marine paints has been also necessary, and phenyl mercuric acetate or its derivatives have been commonly used.

Mercury amplification in food chain

Mercury is regarded to get into the food chain by two routes, water and seed dressings.

It is becoming evident that mercury is concentrated in living tissues. This concentration process has been involving food chains in which small organism containing mercury are ingested in quantity by large organisms, which, in turn, are eaten by still larger one and so on. The organisms further up the chain obtain all the mercury collected by those down the chain. This amplification can give rise to high levels of mercury in the tissues of animals near the top of the chain. Man is at the top of every food chain in which he is involved.

In the case of fish, tiny marine diatoms are having mercury compounds adsorbed to their surfaces. Fish, upon eating these or other small animals or plants, concentrate the mercury in their tissues. Some mercury also get absorbed by fish through their gills from surrounding water. Because of this process, the tissue of the fish has a higher mercury level than the surrounding water. For example, trout exposed to mercury levels of 0.1 ppm for 1 hour per day has been reported to the having a tissue level of 4 ppm of mercury after 10 days exposure.

Some evidence reveals that mercury from seed dressings can undergo translocation in plants and animals. Translocation in plants has been involving the transfer of mercury from seed to crop. In animals, the process makes mercury to move to various parts of the animal or into products obtained from the animal (eggs, for example). It has been possible for mercury to get translocated from grain seed to the harvested grain, then to a hen, and finally to eggs. Data exist to support this idea.

Mercury poisoning and biological methylation

The occurrence of mercury poisoning has been well documented fact. Widespread incidents have been reported only during the last two decades. These incidents occurred due to the eating of mercury-contaminated fish or the eating of seed grain treated with mercury compounds.

Although the toxic action of mercury in the body has been not completely understood, recent research has revealed some information about the problem. The primary facts determined have been as follows:

1. All mercury compounds have been toxic to the body when present in sufficient quantity.
2. Different mercury compounds exhibit somewhat different characteristics of toxicity, distribution, accumulation, and retention times in the body.
3. Certain biological transformations may occur in the environment or the body which change mercury compounds of one type into other types.
4. The ultimate effects of mercury in the body appear to be inhibition of enzyme action and cellular damage caused by the ability of mercury to bind tightly with sulfur containing groups in molecules which are present in both enzymes and cell walls. Such binding is able to inactive enzymes and upsets chemical reactions which they catalyze in the body. The membrane properties of cell walls get disrupted by the binding of mercury, and normal cellular activities are inhibited.

5. The damage done to the body by mercury is generally, usually permanent. There has been no known effective treatment.

Inorganic mercury tends to accumulate in liver and kidney tissue. This causes damage, but also offers a means for rapid elimination through the urinary system. Metallic mercury vapor is having a great capacity for diffusion through the lungs into the blood and then into the brain, where serious central nervous system damage takes place. Normally, inorganic mercury, in the form of compounds, does not remain in the body long enough to accumulate to dangerous levels.

Upon introduction into the body, arylmercurials break down readily into inorganic mercury compounds and behave as indicated above. The compounds PMA, behaves this way.

Alkylmercurial represent the greatest health hazard due to a number of reasons:

1. They easily penetrate into and accumulate in brain tissue. This is accomplished in part since they diffuse easily through biological membranes.
2. They exhibit long retention times in the body. This makes the mercury concentration to build up even though dosages may be low. The estimated half-life of these compounds in the body has been 70 days.
3. Alkylmercurials can be produced from inorganic mercury by the action of certain anaerobic microorganisms. This transformation occurs readily in the mud of rivers and lakes. The extent to which it takes place in the human body is not known, but micro-organisms found in some animal intenstines are able to accomplish the feat.

Methylmercurials appear to be the most dangerous and toxic of the mercury compounds. They can enter the environment in several way-directly, indirectly, intentionally, or unintentionally. A direct and intentional route could be represented by the use of methylmercurials on seed. A direct but unintentional routes has been provided when methylmercurials used in industry or formed as a waste product of industrial processes are dumped into natural waters. The episode at Minamata Bay was caused in this way. An indirect route has been provided when other mercury compounds get methylated (transformed) by biological organisms.

A very serious example of the latter route has been involving inorganic and phenyl mercury from industrial sources. The metallic mercury or compound, upon settling into the bottom mud of lakes, gets converted by anaerobic microorganisms into CH_3Hg^+ or $(CH_3)_2Hg$. This latter compound has been very volatile and escapes from the mud into the water. The $(CH_3)_2$ Hg has been stable in alkaline solutions, but when exposed to acidic surroundings, it changes to CH_3Hg. This ion has been soluble in water and concentrates in living organisms, usually appearing in body lipids. This accumulation occurs due to direct absorption of the ion or an amplication by a food chain process.

In order to prevent environment pollution by mercury, following facts must be kept in mind:

1. The use of all alkyl mercury pesticides must be terminated.
2. The use of all other mercurial pesticides should be restricted to areas of demonstrated need.
3. All industrial users of mercury must be required to reduce discharges of mercury into the environment to levels approximating normal background levels.

The implementation of the above recommendations will not completely solve the problem. Sufficient mercury contamination has been present in the bottom muds of rivers and lakes to continue to yield highly toxic CH_3Hg^+ into the water for years to come. Some sort of decontamination is needed. Large scale pilot experiments on decontamination methods have been postulated:

1. Bottom sediments get covered with fresh, finely divided materials having high absorption capabilities.
2. Bottom sedimets get buried under inorganic, inert materials.
3. The mercury-containing bottom sediments have to be removed by dredging or pumping.

5 Solid Waste Management

At all levels of development human beings produce domestic wastes: at the very minimum these comprise kitchen wastes, ashes from fires, broken utensils and worn-out clothing. In a small agriculture community nature readily accepts these wastes into her natural cycle: animals consume food residues, and other materials are rapidly incorporated into the soil. The industrial revolution in the temperate countries concentrated people in urban areas of very high population density and added new sources of wastes: shops, institutions and factories. Apart from any question of public health it was impossible to accommodate these wastes, arising at a rate of up to one ton/family/year, within the urban areas. Thus, in those countries where industrialization came early, services for the regular removal of domestic and trade wastes have been in operation for a hundred years or more.

Many changes have taken place during this period: the character of the wastes has altered in line with rising living standards, changes in retail distribution methods and fuel technology. The volume has greatly increased; storage methods have evolved from open heaps through portable containers to expendable containers; transport has changed from horses to motor vehicles and from open trucks to compactor vehicles. New problems have arisen—high-rise buildings, supermarkets, industrial wastes of many kinds—and usually efficient solutions have been devised.

At this stage in their development, most industrialized countries exhibit the following common features :

— a high level of national wealth which is still increasing ;

— a taxation system which provides a reliable source of funds for public services such as refuse collection ;

— a long tradition of regular and efficient removal of refuse direct from dwellings and other premises, and usually operated by a municipality ;

— a high level of sensitivity to aesthetic standards within private premises (which does not always extend to standards in public places) ;

— willing acceptance by householders, shopkeepers and others of statutory duties such as providing and maintaining storage containers, sometimes of a specified type and size, and assisting in making containers available for emptying by municipal employees at prescribed times.

During the past twenty years the main thrust of technical development has been towards mechanization, because of rising labour costs, even to the point of installing underground pneumatic transport systems at a few high-rise developments. Although manual emptying of containers continues to be employed almost everywhere, the ratio of vehicles to men is rising and the service is becoming more capital intensive with investment upto Rs. 400000 per collector in some cities.

In the industrialized countries, therefore, basic health and environmental problems have been solved in the storage and collection of solid wastes, although major problems remain in regard to resource recovery and disposal. Despite recurring conflicts over the allocation of national resources as between public and private expenditure, sufficient money is available normally to provide an efficient public service and the citizens cooperate well with the system. The technology of wastes handling is now highly developed; substantial sectors of industry are engaged in the production of equipment; there are established institutions for technical training; there is a wealth of technical literature much of which is devoted to the efficient deployment of labour in relation to capital equipment—another indicator of western pre-occupation with rising labour costs in the service industries.

By contrast, India and many other developing countries are suffering all the problems of urbanization, often with population densities much higher than in any western city, but without the financial resources needed to provide solutions of the kind used in the West.

The developing countries are not unaware of the importance of avoiding the environmental pollution which shadowed urbanization and industrialization in europe and north America. The quality of urban environment is a matter of growing concern to them and the importance of efficient solid wastes management is increasingly recognized. The West has often been able to provide developing countries with technical guidance in environmental matters, but solid wastes may prove to be an exception; there are too many climatic, economic and social differences for systems to be successfully transplanted. Even the technical literature of the West may be of little value of either training or operational guidance.

Sources and characteristics

"Solid wastes" is the term now used internationally to describe non-liquid waste materials arising from domestic, trade, commercial, industrial, agriculture and mining activities and from the public services. "Non-liquid" is a relative term because sludges of certain kinds fall within the scope of solid wastes management; these arise primarily from industrial sources and from sewage treatment plants.

Solid wastes comprise countless different materials; dust, food wastes, packaging in the form of paper, metals, plastics or glass, discarded clothing and furnishing, garden wastes and hazardous and radioactive wastes.

Domestic wastes have three main characteristics: weight generated, density and constituents; these vary not only from country to country, but from town to town within a country, according to the level of industrial development and other factors :

— the range of weight generated/person/day usually lies between 250 and 1000 g. worldwide.
— density varies from 100 kg/cu. metre to 600 kg/cu. metre.
— thus volume may range between 1/2 litre and 10 litres/person/day.

The main constituents of domestic wastes vary as follows :

— vegetable/putrescible matter, 20% to 75%,
— Inter metals, 5% to 40%,
— paper, 2% to 66%.
— glass, 0% to 10%,
— metals, 0% to 15%.

No rational decisions on solid wastes systems are possible until data of this kind are available. The method and capacity of storage, the correct type of collection vehicle, the optimum size of crew and the frequency of collection depend mainly on volume and density; climate also has some influence. The disposal method may be conditioned by the proportions of saleable materials that could be recycled, the vegetable content which may be compostable or the paper and plastics content which could be a source of energy.

Just as solid wastes comprise a vast number of materials, they arise from a multitude of separate sources; a city of a million inhabitants may have a quarter of a million separate sources, as well as many kilometers of streets upon which solid wastes accumulate. Thus the four main aspects of solid wastes management are :

— storage at or near the point of generation
— collection,
— street cleansing,
— disposal.

An apt definition of solid wastes is "matter in the wrong place," implying that a material becomes waste only when a specific owner ceases to have a use for it. Yesterday's newspaper is waste to the man who bought it, but it could be the raw material for a paper mill. Current waste disposal philosophy is to endeavour to treat all wastes as resource materials: some for recycling; some for conversion to fertilizer or as a source of energy; and the balance for land reclamation.

Health and environmental implications

There are potential risks to health and to the environment from improper handling of solid wastes. Direct health risks concern mainly the workers in this field, who need to be protected, as far as possible, from skin contact with wastes. (The National Environmental Engineering Research Institute, India has carried out studies that show a higher incidence of intestinal parasites in collectors of domestic wastes than in the population as a whole).

There are also specific risks in holding wastes from hospitals and clinics. For the general public, the main risks to health are indirect and arise from the breeding of disease vectors, primarily flies and rats. Improper storage and disposal provide the conditions under which these risks arise.

The most obvious environmental damage caused by solid wastes in the countryside is by uncontrolled dumping of city wastes. More serious, however, and often unrecognized, is the transfer of pollution to water, which occurs when the leachate from a refuse dump enters surface water or wastes, either in the open air, or in plants that lack effective treatment facilities for the gaseous effluents.

Industrialization introduces dangers of a different kind: hazardous wastes from industry present risks during transport and disposal. Traffic accidents can result from wastes: they have caused death and injury to people in the vicinity. Improper disposal of such wastes has resulted in the death of humans and animals through contamination of crops or water supplies.

Finally, there is the specific danger of the concentration of heavy metals in the food chain, a problem that illustrates the relationship between solid and other wastes. It has sometimes happened that liquid industrial effluents containing heavy metals have been discharged to a drainage system and have contaminated the sludge has leaving the treatment plant. These metals can be taken up by the plants growing on land on which sludge has been deposited, creating risks to the animals which graze and the humans who consume these animals.

Economic implications

Labour and transport absorb the major part of the operating cost of solid wastes management services. It is usual for a city to employ between two and five workers for each 1000 population. Three workers/1000 population may represent about 1% of the total national workforce. Transport is required on a scale of about 1 heavy vehicle/ 15000 population. Thus, solid wastes management services can absorb upto 1% of the gross national product, and are one of the most expensive city services. Efficiency of systems and high labour productivity, therefore, are of vital importance.

The level of mechanization that should be adopted for solid wastes management systems relates directly to the cost of labour, as compared to that of plant and energy. There is not much variation, worldwide, in energy or mechanical plant costs, but there is an astonishing range of labour costs.

Thus there are no universally applicable solid wastes management systems: every country must evolve and indigenous technology based on the quantity and character of the wastes, the level of national wealth, wage rates, equipment manufacturing capacity, energy costs and the availability of foreign exchange for the purchase of imported plant.

Human resources

A critical factor in the evaluation of alternative systems of solid wastes management and the establishment of efficient organization is the quality of management. It is necessary to deploy a complete set of technical skills which derive from several professional disciplines. These include civil and mechanical engineering, chemical engineering, transport organization, land use planning and economics. Thus, solid wastes management is a team work, and whatever the original discipline of the top manager, his prime qualification should be his specialized background in this field. A second requirement, no less important, is skilled "man-management," the purpose of which is to obtain high productivity from a contented staff at all levels.

Refuse characteristics

The generation of domestic and trade wastes is a remorseless and continuing process. Throughout the day a housewife is performing such tasks as sweeping floors and trimming vegetables: the shopkeeper is discarding packing materials and spoiled goods: the shoemaker is paring leather; almost every member of the population is continuously creating wastes of one kind or another. In the case of human body wastes, a sewage system provides facilities for transfer from the production point to the disposal place by a continuous flow process in pipes containing water. Despite the successful use of pneumatic transfer in underground pipes at a few residential developments in Europe, no such methods is generally practicable for solid wastes. Thus refuse collection must be a batch process, or a series of batch processes, whereby wastes are stored at the point of origin for a given period, before being transferred to a vehicle, the contents of which represent one batch of many arriving each day at the disposal point.

The main sources of those solid wastes for which a municipality normally assures responsibility are domestic premises, shops, offices, hotel, institutions, and small factories, together with refuse swept from the streets. Domestic wastes often account for about 75% of the total.

The main constituents of solid wastes are similar throughout the world, but the proportions vary widely from country to country and even within a city, because the variations are closely related to income levels.

As personal income rises, paper increases, kitchen

wastes decline, metals and glass increase, total weight generated rises and the density of the wastes declines. This will be more clear from the table given hereafter.

There are often local variations in wastes generation and constituent proportions over weekly and seasonal cycles; the former is related to the pattern of work and leisure and the latter to seasonal food products and sometimes to fuel residues arising from space heating in winter.

Item	Indian city	N. European city
Paper	4%.	27%
Vegetable/pursuable matter	75%	30%
Dust etc., under 10mm	12%	16%
Metals	0.4%	7%
Glass	0.4%	11%
Textiles	3%	3%
Plastics	0.7%	3%
Others, stones, ceramics etc.	7%	3%
Weight/person/day	414g	845g
Weight/dwelling/day	2.5 kg (6 persons)	2.5 kg (3)
Density kg/cu.m	570	132

Clearly, the amount of work involved in refuse collection depends upon the weight and volume of wastes generated and the number of collection points from which the wastes have to be removed.

The extent to which these factors may vary is shown in the following table: The following important issues emerge from these figures :

— The importance of density determining the volume to be stored and collected: about 4 litres from an Indian dwelling with a daily collection compared with about 130 litres from a European dwelling (many houses have two 70-litre bins) for a weekly collection;

Item	Indian city	N. European city
Generation		
Volume/person/day	0.72 litres	6.4 litres
Volume/person/day	(6) 4.32 litres	(3) 19.2 litres
Sources/1000 population	166	333
Collection points : (per million population)		
House-to house collection daily	1,66,000	
House-to house weekly, per day	56,000	1,11,000
Communal storage points/million population		56,000
(for average contents 1 cu.m)		
Points to be visited daily	720	

— the higher rate of occupancy in India helps to reduce the number of sources, but this is far outweighed by the need for more frequent collection,

— the use of the communal container can reduce the total number of collection points to only 0.5% of the total sources; this enormously simplifies the organization of the work, but may cause serious environmental problems.

Work components of refuse collection

A refuse collection service requires vehicles and labour. For their efficient development, it is necessary to have a clear understanding of the three main components of refuse collection ;

- travel to and from the work area;
- the collection process, which includes: transfer of the wastes from storage to the vehicle and travel between successive collection points;
- the delivery process whereby the contents of the vehicle are transported to the disposal site.

During non-working hours vehicles will be kept at a depot with enclosed parking space. The distance between the depot and the collection area should be kept to a minimum because time spent travelling is unproductive. In the case of motor vehicles this requirements may have to be balanced against the need to centralize facilities for maintenance and stores, particularly fuel supplies, and to centralize the allocation and control of drivers and vehicles.

The slower speed of animal carts and handcarts requires the provision of closely spaced district depots, and these are also much efficient for the control of the collectors, who can book on and off within a short distance of the area where they will be working.

The many methods of transferring wastes from storage to a collection vehicle fall into three main categories.

- the direct emptying of a portable storage container into the vehicle, normally used when the vehicle can be positioned close to the container ;
- the emptying of a portable storage container into a transfer container (usually a large basket) which is then emptied into the vehicle, normally used when the storage container is some distance from the vehicle, in order to avoid the return walk with empty container.

— the transfer of wastes stored at ground level, which usually requires that the wastes be raked or shovelled into the vehicle.

These three categories have been set out in ascending order of work content. Thus the first is the most efficient in terms of labour and vehicle productivity; it is also the method minimizing human contact with the wastes.

Travel between successive collection depends first upon the spacing. When these are some distance apart, as with large communal storage sites, travel by motor vehicle, although it represents lost time, will be at normal road speed and the collectors will ride on the vehicle. It is thus an efficient method of transporting the workers between sites.

When collection is from house to house, however, the collectors will walk the short distances between containers and the vehicle will move slowly and at intervals. For this element of travel the motor vehicle is very ineffectively employed; it incurs heavy transmission wear and high fuel consumption. Handcarts and animal carts are much more efficient: they operate at their optimum speed and no energy is used while they are stationary.

For collectors walking from house to house, the distance to be walked is proportional to the number of men in a team. One man alone walks from one house to the next, in a three-man team each man collects from every third house, thus labour productivity declines as team size increase. Vehicle productivity, however, increases with team size as it is loaded more quickly.

The delivery process, when a full vehicle travels at normal road speed from the collection area to the disposal site, represents maximum productivity for the vehicle, but lost time for the collectors if they accompany it. Handcarts and animal carts are very inefficient for this operation because of their slow speeds and limited capacity. The following conclusions can now be drawn :

— Minimum physical infrastructure requirements are a central workshop with parking space for motor vehicles and district depots for mustering and controlling collectors, handcarts and animal carts; the locations of depots should minimize travel time between depot and working area;

— systems which provide for the direct emptying of portable storage containers into a vehicle offer the highest productivity and the lowest health risks to workers;

— for the collection wastes direct from houses, large teams give low labour productivity and higher vehicle productivity;

— hand carts and animals carts may be more efficient than motor vehicles for the house-to-house collection phase;

— motor vehicles are the most efficient for transport of complete loads from the collection area to the disposal site.

Four basis collection systems have been involved in relation to the amount of work imposed upon the householder :

— communal storage which may require delivery of the wastes by the house holder over a considerable distance;

— block collection, where the house holder delivers the wastes to the vehicle at the time of collection;

— kerbside collection, where the householder puts out and later retrieves the bin;

— door-to-door collection where the collector enters the premises and the householder is not involved in the collection process.

Collection from communal sites

The organizations of refuse collection is greatly simplified by the use of large communal storage sites. The city of Delhi, for example, is served by about 1250 sites having capacities from 0.5 to 10 tons for a population which exceeds 4 million. Delhi probably has about 800000 separate sources of wastes (dwellings and shops) and a frequent collection direct possibly greater expenditure. However, when sites are widely spaced, a great deal of domestic wastes are deposited in the streets by householders too lazy to carry it to the depot or masonry enclosure. It is significant, therefore, that only by employing about 10000 sweepers is Delhi kept in a tolerable standard of cleanliness.

While the use of large communal sites appear to be fairly cheap and simple solution, it may transfer much of the burden of refuse collection on to the street cleansing service and actually increase total costs, because it is cheaper to collect refuse direct from a house than to sweep it up from the streets. In Delhi, in the year 1974 Rs. 31,000,000 were spent on sweeping, and only Rs. 7,330,000 on refuse collection.

The use of large, widely spaced communal storage sites is usually a failure because the demand placed on the householder goes beyond his willingness to cooperate. Communal storage points should, therefore, be at frequent intervals, Madras, Bangalore and Manila provide fixed concrete receptacles with capacities between 100 and 500 litres on footways or verges at intervals of 50 to 200 metres. The objections to these have been stressed earlier, but they are fairly successful because they place reasonable and acceptable duty on the residents, thus very little domestic waste is thrown in the street.

Both the large masonry enclosure and the smaller

concrete bin are inefficient in the use of manpower and vehicles. Wastes have to be removed by rake or shovel and basket; it is a slow process and vehicle waiting time during the loading process is excessive. Vehicles employed on this work tend to interfere with other traffic in the street. The following work performance have been recorded :

Delhi, Masonry enclosures	1.4 tons/man/day
	7 tons/vehicle/day
Bangalore, concrete pipes	1.2 tons/man/day
	6 tons/vehicle/day

Drums of 200 litre capacity are far from an ideal solution, but the fact that two men can usually empty them directly into a vehicle with a low load line greatly increase daily performance; probably to about 5 tons/man/day and 10 tons/vehicle/day.

Block collection

In this system, a collection vehicle travels a regular route at prescribed intervals, usually every two days or every three days, and it stops at every street intersection, where a bell is rung. At this signal the residents of all the streets leading from that intersection bring their wastes containers to the vehicle and hand them to the crew to be emptied. A crew of one or two men is adequate in number as they do not need to leave the vehicle.

Block collection should be operated frequently, as otherwise the weight of wastes to be carried to the vehicle may be beyond the capacity of some of the residents. It has a significant advantages over a kerbside collection in that bins are not left out on the street for long periods. This system is operated in Mexico City and during a study carried out by Solid Wastes Division of the Sub-Secretariat for Environmental Protection the following data were collected. They represent the average of four routes :

Collection route, total length	2.74 Km
Number of stops on route	25
Average distance between stops	110 metres
Bins emptied per stop	33
Total bins emptied on route	858
Average load, total weight	3653 kg
Dwelling on route	841
Average weight/dwelling	4.34 kg
Total time for load	2 hours 27 min.
Average travel time point to point	1.15 min.
Average loading time per point	7.10 min.

The daily performance achieved by this system is about 3.5 tons/man/day and 7.0 tons/vehicle/day.

Kerbside collection

This, like the block collection, requires a regular service and a fairly precise time table. Residents must place their bins on the footway in advance of the collection time and remove them after they have been emptied. It is very important that bins of a standard type should be used, as otherwise it is likely that wastes will be put out in improvised containers, such as cardboard boxes or even in loose heaps; when this occurs some of the waste are inevitably scattered by animals and wind, thus increasing the work of street cleaning. Kerbside collection is never entirely satisfactory. Problems include :

— bins sorted by scavengers,
— bins stolen,
— bins turned over by goats or cattle,

— failure of the householder to retrieve the bin quickly. The worst example of this last problem is when the bins are kept permanently on the footway; this is not uncommon in high income residential areas of certain Asian cities.

However, kerbside collection is unavoidable with some types of house construction, and it is the cheapest method of house-to-house collection. When the rate of wastes generation is high and collection infrequent, very high labour productivity can be achieved. For example, in one city in USA a one-man crew collects up to 10 tons/day-400 dwellings at an average of 25 kg/dwelling. In all developing countries, however, the wage-rate to vehicle-cost ratio would be much less than in USA and it would be profitable to employ a crew of at least four men. Productivity would also be reduced as the weight per dwelling would be much less.

Door-to-door collection

This is the system in which the householder does no work: the collector enters the garden or courtyard, cares the bin to the vehicle, empties it, and returns it to its usual place. It is costly in labour because of the high proportion of working time spent walking in and out of premises and from one dwelling to the next, but is the only really satisfactory system. A study in USA showed that this system costs about twice that of kerbside collection, but this ratio would be greatly reduced in countries where labour cost is low.

The problem with door-to-door collection in developing countries is that vehicle productivity would be very much less than in Europe or USA if collection was at high frequency. Since high vehicle productivity must be the main aim of developing countries, door-to-door collection by the conventional western method of heavy motor vehicle and crew is very unlikely to be a viable system unless the interval between collections is extended to a week. This is unlikely to be acceptable in tropical countries as a principle and in any case the majority of dwelling in most cities need a daily collection (or communal storage).

Sanitary Landfill Disposal

Land disposal, or more accurately, the burying of wastes, is the only approved method of disposal which is performed at a single site. Incineration, composting, and salvage are either a form of refuse handling or processing. They are not complete methods of disposal, and they require disposal of residue or the removal for sale or internal consumption of compost.

Land filling is not really a self-contained method of disposal, because matter can be leached by water moving through the system. When rainfall infiltrates the landfill surface or when the ground water table saturates a part of the landfill, a return to equilibrium moisture conditions requires that water in excess of field capacity will drain from the landfill. As the water (leachate) drains from the landfill, it carries mineral and organic matter in suspension or as climate and geology of each landfill. In some geological formations water drainage is primarily vertical, and unless a nearby ground water source is affected, it is difficult to horizontally trace pollution from sanitary landfills. In other formations the impervious layers interrupt the vertical drainage and convey the leachate horizontally until the impervious aquifer outcrops and causes contamination of surface waters.

Sanitary landfill can be defined very simply as the use of solid wastes for land-reclamation, a typical example being the restoration, by filling to the original level, of man made surface dereliction such as a disussed surface mineral excavation. Solid wastes may also be used to improve natural features by raising the level of low-lying land to enable it to be used or cultivation or industrial development. Described in this way, sanitary landfilling has the virtue of being a method of refuse disposal which confers environmental improvement by restoring dereliction or improving natural contours.

Most solid wastes however, are very offensive materials which provide an attractive habitat for such disease vectors as flies and rodents, and which pass through a lengthy process of decomposition the products of which can cause serious water pollution. Thus sanitary landfilling has two essential features which differentiate it from crude dumping:

— only sites that will be improved, not degraded, by a change of level, are selected;

— simple engineering techniques are used to control the manner in which the wastes are deposited, so that dangers to public health and the environment are avoided.

The methods of control include two classical requirements:

— wastes must be deposited and levelled in layers not exceeding about 2 metres in depth;

— the whole of the surface of each layer of wastes is covered with soil or other suitable material to a depth of 15-25 cm. on the same day as the wastes are delivered to the site.

Limiting the depth of a layer is required partly because of the instability of newly deposited wastes over which vehicles must pass; also to limit the fire risks which arise from various causes. Covering the wastes serves several purpose: it is no longer assessable to flies; attempts by rats to penetrate the covering material are immediately apparent; odorous are avoided; and most important, the heat resulting from decomposition of organic mater is retained, and assists in the destruction of fly larvae and pathogenic organisms in the wastes.

Sanitary landfill is usually by far the cheapest method of refuse disposal. Because of this, and its comparative simplicity of operation, one would expect that it would have been adopted the world over, except for the rare case of a city having no suitable site within reasonable distance, but this is not the case.

Unfortunately most of the world's wastes are disposed of by uncontrolled dumping which blights the land for any future use and causes serious risks of water pollution and vector breeding. In some cases, wastes are levelled by a bulldozer so that ultimately the land can be used for another purpose, but this does not prevent pollution occurring throughout the filling operations and for some time afterwards. Very few cities operate sanitary landfilling to standards which totally control health and environmental dangers; most of those that do are in the industrialized countries.

The record of the wealthier nations, however, is by no means a good one. Numerous landfill sites have been so badly operated as to cause devaluation of property in their vicinity, and bitter criticism from the public.

"Bad practices of the past have often created a climate of public opinion that leads property owners to oppose the location of any kind of treatment plant or disposal site in their vicinity, and renders the acquisition of sites extremely difficult, even when high standards of operation are promised. Even professional urban planners are often reluctant to allocate land for solid wastes disposal."

Paradoxically, many municipalities in the industrialized countries which avoided paying that extra Rs. 10 ton needed do provide good management and operation of a sanitary landfill site, have ultimately found themselves paying Rs. 150 ton for incineration, not through any lack of suitable landfill sites, but because public opinion has been outraged by poor landfill standards.

The record of the developing countries is even worse; crude dumping is still the most common method of disposal and causes the following hazards :

— fly generation
— encouragement of rodents
— static water pollution and aerial nuisance
— river pollution
— sea pollution
— fire and smoke pollution

Decomposition in sanitary landfill

Decomposition in a sanitary landfill arises from chemical changes which are brought about by bacteria which are present in the organic content of the wastes when they are collected, and which multiply rapidly, particularly when the wastes are effectively sealed. There are three main types of bacteria which are found in a landfill: the saprogenic, the agent of decomposition; the zymogenic, which are agents of fermentation (they are used in commercial processes such as brewing); and the pathogenic, which are associated with human diseases.

Both the saprogenic and the zymogenic bacteria are found in form which vary according to the supply of oxygen. There are aerobic types for which oxygen is a necessity. There are obligate anaerobes for which oxygen is a disadvantage, and facultative anaerobes which can adopt themselves to varying supplies of oxygen.

During the first stage of decomposition, aerobic bacteria are dominant, but as the available oxygen declines their work is taken over by anaerobic bacteria; facultative anaerobes will be present at all times.

The bacteria have the ability to create certain colonial solutions known as enzymes which have the power to breakup stable compounds. The enzymes are used by the bacteria to reduce organic matter to a form of nourishment for them and it is this process of breaking down, and making a soluble, relative stable compounds that we know as decomposition.

The consequences of all these complex changes include:

— temperature changes within the landfill;
— the production of gases: hydrogen, oxygen, nitrogen, carbon dioxide, methane, and sulphuretted hydrogen;
— the production of soluble and particulate decomposition products which can be carried away from the site as a leachate when water passes through the landfill.

Temperature— Irrespective of ambient temperature, the internal temperature of a landfill rises soon after the deposit of the wastes as a result of heat emitted by the activities of aerobic bacteria.

The main significance of this rise in temperature, is possible significance in the destruction of pathogens. The conclusions reached by two British Investigators, Jones and Owen ("Some Notes on the Scientific Aspects of Controlled Tippin." City of Manchester, 1934) were as follows :

"—it is highly improbable that pathogenic bacteria can multiply and continue to live in a tip." The factors to be taken into account when considering the likelihood of a danger arising from pathogenic bacteria are as follows :

"The temperature inside a tip rises during the early period of decomposition to a point, in most cases, higher than the thermal death points of pathogenic germs."

"Pathogenic germs are difficult to keep alive even in the laboratory."

"The saprogenic organisms will tend to outnumber and exterminate them."

"Unless the organisms are spore formers they will perist rapidly under ordinary conditions of earth burial; all the common pathogenic bacteria likely to be found in a controlled tip are no-spore formers."

Gases—Four gases are of significance: sulphuretted hydrogen, carbon monoxide, carbon dioxide and methane. Sulphuretted hydrogen is toxic, but only small traces have been found, and any large concentration would give ample warning of its presence by its extremely offensive odour. Carbon monoxide, which is present in large amounts, is diffused very slowly and offers no risk on the surface; in excavations on sanitary landfills it has been found at a level of one percent which could cause unconsciousness and death if a person were to be trapped.

Because methane is diffused very slowly from the site, in concentrations very much below the level at which ignition would occur, it rarely presents any risk. Conditions have arisen, however, where methane has seeped through fissures in the soil into buildings adjoining the landfill site; in such cases a dangerous concentration could occur.

Leachate—Organic and inorganic matter present in crude refuse, and the products of decomposition, can be leached by water; either rainfall or ground water, passing through the wastes. This is a potential cause of serious water pollution. In a series of experiments it was found that a layer of refuse 1.5 metres deep, compacted and covered with soil, would absorb rainfall for several weeks and thereafter produce a leachate equivalent to about one third the annual rainfall in that area, which was 635 mm. This leachate was 20-30 times as strong as settled sewage, having a BOD of 6000-7000 mg/litre, and contained organic carbon, ammoniacal and organic nitrogen, ammonia, chloride and sulphate. The leachate continued for several years, but within less than three years pollutants had declined to a very low level except for the sulphate.

It follows that wastes should not be deposited in such a way that water can pass through them to a stream, not to an underground water supply.

Control of hazards

There are a number of hazards—such as pathogenic organisms, insects, rodents, birds, airborne litter, fire and water pollution—which have to be controlled in a sanitary landfill.

Pathogenic organisms—Control over pathogens is dependent upon a rigorous policy of covering the wastes, soon after deposit, with a layer of soil or other suitable material at least 15 cm. in thickness. This serves both to isolate the wastes and to retain the heat which is quickly generated during aerobic decomposition.

It is probable, however, that workers at the site will be exposed to the wastes before they are covered. In developing countries the risks to workers by skin contact with wastes may be greater than in the industrialized countries because street sweepings may contain human factual matter. Protective clothing is necessary, therefore, and should include rubber boots, gloves and overalls which are changed and washed regularly. The provision of personal hygiene facilities at the site is also important.

Insects—The main source of insects will be the eggs of flies which have been deposited in the wastes before they arrive at the site. Most of these will be buried deep in the wastes and will succumb to the temperature increase. Near the surface, however, the rise in temperature will not be so great and the eggs will hatch into larvae; only the provision of an adequate layer of suitable covering material, well compacted, will prevent their emergence and pupation. It is the importance of providing an effective barrier of this kind which has caused the British Government to recommend a covering thickness of 25 cm.

Another obvious advantage of covering is that it denies access to the wastes to flies that are already at the site and thus avoids a major problem of open dumping: continuous cycles of breeding in exposed wastes at the site.

One of the recommended precaution in sanitary landfill is not to permit cans and motor types to lie about on the surface. They not only spoil the appearance but when filled with water they encourage the breeding of mosquitoes. Depressions in the surface of the site which permit ponding after rain should be avoided for the same reason.

If for any reason these methods of control are not adequate it will be necessary to have recourse to insecticides. They should be used daily as long as is necessary. The rate of application recommended in "Sanitary Landfill Design and Operation" (US, EPA. 1972) is as follows:

Insecticide	*Approximate rate of application (g/ha)*
Malathion	672
Dischlorvos	336
Naled	111-222
Dimethoate	111-222
Diazinon	336
Fenthion	450
Ronnel	450

It is the writers, experience that the need for insecticides arises only because of some failure of the normal control measures such as allowing a large surface area of wastes to remain uncovered, or the use of a very porous or thin cover which allows larvae to reach the surface.

Rodents—Rodents may be attracted from surrounding farms, or they may be delivered to the site in a load of wastes. A clean, unbroken and relatively smooth surface of covering material allows a burrow to be instantly detected and recognized. Regular inspection should be made and if the beginning of rat infestation is observed, a baiting programme should be carried out by an experience exterminator. It is normal practice to use an anti-coagulant poison.

Airborne dust—During long dry periods the surface of a sanitary landfill can become very dusty, causing unpleasant working conditions for the men, and excessive wear of vehicles and plant. Watering carts are frequently used to reduce this nuisance; and waste oils are suggested in "Sanitary Landfill Design and Operation."

"Periodic treatment or multiple & praying at a rate of 1-4 litres per sp. metre may be necessary. After several

treatment, a packed, oily, soil crust usually develops that has good resistance to traffic abrasion and is moderately resistant to water. Good penetration by the oil can be expected in more permeable soils. Clayey soils or tightly knit surface may resist penetration, in which case it may be desirable to scarify the surface lightly, apply-oil and compact the surface."

Airborne litter—Airborne litter is often a major nuisance in the industrialized countries, and during gales at desert sites in countries having a significant paper and plastic content in wastes. In the latter case environmental degradation of this kind has been observed extending over many sq. km. It is helpful to keep the size of the working face to a minimum, but this does not prevent litter being carried away during the period when the wastes are being discharged from a vehicle. It is necessary, therefore, to employ a movable barrier of perforated screens to prevent litter being blown off the site.

Convenient dimensions for a screen are 21/2 metres long and about the same in height. The frame can be made from timber or light steel angle. To enable the screen to stand unsupported horizontal transverse members are attached by brackets to each bottom corner; these should be about 1 metre long. The whole area of the screen is covered with wire netting (chicken wire) of 20 mm to 40 mm gauge. Enough screens are necessary to form and are sufficiently long to catch everything blown from the working area. They must be cleaned regularly if they are to remain effective, and in very high winds they may require to temporarily supported by guy ropes secured to pegs driven into the ground.

Fire—Fire at a sanitary landfill can arise from innumerable causes: hot ashes in a vehicle delivering wastes; a cigarette thrown by a worker; the sun's ray though a fragment of glass on the surface. With some kinds of wastes the consequence of fire may be very serious and underground fires have been known that ultimately caused the collapse of the surface into voids caused by the fire. The high density, high vegetable, wastes of developing countries are unlikely to give rise to risks of this kind but is important to try to avoid fires and to be prepared to deal with them if they arise. Workers should not be allowed to start fires to dispose of bulky wastes such as tree loppings, and private scavengers, who may use open fires for cooking, should be prevented from obtaining access, especially outside normal working hours.

It is useful to have a supply of water on the site: the water cart used for laying dust could serve a double purpose. Once a fire takes hold, however, it is usually necessary to dig out the material which is burning before water can be applied effectively. In the early stages this can be done by long rakes but once a fire becomes deep seated the excavation of a large area by bulldozer may become necessary.

The wastes of tropical and developing countries have often been seen on fire, but the fires have invariably been on the surface and have never presented the problems of control that occur in Europe where wastes sometimes contain a high proportion of solid fuel such as cinder and coal fragments.

Birds—Birds are normally attracted to landfill sites and may cause risk to aircraft it there is an airport in the vicinity. This problem is exacerbated. If the birds have routine flight path between the sanitary landfill site and another source of attraction, such as a body of water, and their flight path crosses the approaches to a landing ground. Many methods of discouraging birds have been tried, from cannon to falcons. But none has been entirely successful.

This problem is one that does not appear to be controllable and the solution lies in the selection of sites which avoid creating the risk of birds-strikes on aircraft.

Surface-water pollution—The pollution of static water, ditches, river or the sea occur when a sanitary landfill adjoins a body of water. The normal source of the leachate causing this pollution is rain falling on the surface on the fill, percolating through it and passing over an impermeable base to water at a lower level. Only a proportion of total precipitation emerges as leachate; some is lost by evaporation and transpiration. In the British experiments of 1961 an annual rainfall of 635 mm produced a leachate equivalent to 200 mm. In arid areas there may be insufficient rainfall to create a leachate, but in all other cases the production of leachate from rainfall is unavoidable during the early stages of filling. The quantity of leachate can be substantially increased when upland water drains across the sites of the landfill, but the worst case is when a stream crosses the site. The solutions to these problems lie in appropriate site engineering such as :

— diversion or culverting of all watercourses which flow across the site,

— diversion of upland water by means of drainage ditches along appropriate contours,

— Containment of leachate arising from precipitation by the construction of an impermeable barrier where necessary, such as a clay embankment adjoining a river,

— Grading the final level of the site so that part of the precipitation is drained across the surface, thereby reducing percolation below the level needed to produce a leachate.

Work of this nature will obviously add to the cost of a sanitary landfill project, but it is usual to find that when capital expenditure is spread over the life of the project, the cost/ton is still very much cheaper than for any alternative method of disposal. Furthermore, some of these forms of expenditure, such as culverts of river walls, represent capital assets of continuing value when the reclaimed land is handed over for its final use, perhaps agriculture or recreation.

Groundwater pollution every proposed landfill site should be the subject of survey and evaluation by a hydrogeologist, and where there is a risk of leachate passing directly to under ground water which is the source of a public supply or to wells, the site must be rejected. The following comments are designed only to indicate the nature of this problem and do not pre-empt the judgment of the local hydrogeologist.

Potential sites can be divided into three main groups:

— impermeable sites,

— sites on which subsoils such as sand and gravel may the provided a measure of filtration,

— fissured sites from which a leachate may be able to travel many kilometres.

The classic impermeable site is a clay pit from which there are no pollution risks because the leachate cannot escape.

On sites which provide filtration the British Technical Committee reported as follows :

"Experimental filtration at a rate of about 30 cm. per day of the polluted percolates through anaerobic filters of graded sand and gravel produced remarkable results. In a distance of about 7.3 metres practically all bacterial pollution and most of the organic pollution was removed. This was confirmed in full scale tipping..."

The experiments...indicated that sub-soils such as sand and gravel would act as an effective filter for bacterial and organic contaminates. For such contaminates as chloride and sulphate, reliance would mainly have to be placed on dilution but generally it is considered this will be sufficient in many cases when the quantity of water abstracted from an underground source of water supply though to be at risk is related to the quantity of percolate..."

On this question the following observations are made in "Sanitary Land Fill Design and Operation" (EPA, 1972):

It is a basic premise that groundwater and the deposited solid waste not be allowed to interact. It is unwise to assume that a leachate will be diluted in a groundwater because very little mixing occurs in an aquifer since the ground water flow there is usually laminar.

"When issuing permits... many States require that groundwater and deposited solid wastes be 0.6 to 10 metres apart. Generally 1.5 metre separation will remove enough readily decomposed organic and coliform bacteria to make the liquid bacteriologically safe. On the other hand, mineral pollutants can travel long distances through soil or rock formation. In addition to their considerations, the sanitary landfill designer must evaluate the: (1) current and projected use of the water resources of the area; (2) effect of the leachate on groundwater quality; (3) direction of groundwater movement; (4) inter-relationship of this aquifer with other aquifers and surface water."

Where there is doubt as to the characteristics of a proposed site the British Technical Committee suggests the following precautions:

(a) Water proofing the base prior to tipping, i.e. with puddled clay (or chalk), collection of the percolate and its disposal...

(b) Rapid tipping of successive layers of refuse in order to utilize its capacity to absorb rainfall and to cover the final layer with a waterproof earth cover with a good cross-fall to provide a ready run-off of rainfall."

Similar advice is given in "Sanitary Landfill Design and Operation:"

"An impermeable linear may be used to control the movement of fluids. One of the most commonly used is a well-compacted natural clay soil, usually constructed as a membrane 30 to 90 cm. thick. It must, however, be kept moist.

"The use of an impermeable barrier requires that some method be provided for removal of the contained fluid."

Many developing countries may find the collection and treatment of leachate a difficult and costly procedure and one which has no residual value in relation to the ultimate use of the land.

The recommended approach is to reject sites at which problems are revealed by hydrogeological survey. The major pollution problem in the developing countries appears to be the contamination of surface water by run-off and this can be controlled by relatively simple site engineering.

Incineration

Open burning, barrel burning, and other related uncontrolled forms of burning have a long history of use. Industry and farmers each have exploited the combustibility of their waste products. In domestic refuse the dominant fractions of paper and food wastes are burnable, as well as other constituents such as wood. Many liquid wastes and pathological wastes are best disposed of by incineration.

Originally, solid waste incineration was practised to reduce the quantity of refuse or disposal. After it was proven that heat could destroy most pathogens, incinerators were used in hospitals for destruction of pathological wastes. Incinerators to burn sludge from the treatment of modestic sewage are used for volume reduction and pathogen destruction. Today, in spite of our advanced knowledge of combustion, the design of refuse incinerators is lagging in technology, especially in terms of air pollution control, organic destruction and multi-purpose use.

Multipurpose use can mean integrated liquid and solid disposal systems or comprehensive regional plants for industrial and domestic refuse disposal. Heat recovery to produce steam for heating or power generation is the multipurpose use most frequently proposed. Power recovery units have been built in the United States at Ocean side, Long island for operation in conjunction with desalinization and at the Norkfolk Navy Yard. As the cost of incineration increases, a greater emphasis on multipurpose use will have to be instituted to keep incineration economically competitive.

With few exceptions, incinerators are not "good neighbours," and the environmental nuisances of dust, noise and air pollution have provoked communities to an anticinerator philosophy. To overcome this negative community feeling is going to require that incineration prove its worth and that imagination be used in the design of future units.

Incineration of solid wastes yields the highest percent of volume reduction except for pyrolysis. Again, except for pyrolysis, the weight reduction achieved in incineration is greater than for other disposal methods. The weight reduction achieved depends on the amount of glass, metal, and dirt which is not combustible. Generally, the weight reduction is between 50 and 80% and volume reduction between 70% and 85%. It is this reduction in waste weight and volume which gives incineration its most appeal as a processing technique.

Unlike a sanitary landfill, incineraton of solid wastes can be performed on the premises of apartments, supermarkets, departments stores, and similar establishments. The size of these on-site units begins at about 50 lb/hour, and the upper limit is 1-2 tons/hour. On-site incineration reduces the weight and volume of wastes which must be removed for final disposal.

Composting

Composting involves the biological stabilization of solid matter either under aerobic or anaerobic conditions. The end product of composing is an organic material which could have beneficial value as a soil conditioner or plant mulch. Selected nutrients could be added in order to assure that the end product has minimum concentrations of nitrogen, potassium and phosphorus for fertilizer value. If the material does contain a reproducible average concentration of nutrients, then compost can be legally marked as an organic fertilizer.

Existing methods of composing parallel the historical development of composing techniques. The basic techniques used in the United States are window composting and compost digesting. The compost digestors can be either aerobic or anaerobic, batch or continuous. The most common process methods in the United States are the continuous aerobic type and are generally thermophilic, i.e; the dominant organisms are most efficient at temperatures as high as 170°F.

In addition to producing a modified solid waste material which can be useful in land reclamation, composting does yield a volume reduction of solid waste by about 40-60% of the compostable fraction.

During the short history of commercial composting in America, it has been assumed that composting of solid wastes is a method of conservation which produces a salable and useful end product. Marketing the final product has been one major drawback to wide scale use of composting as a method of solid waste disposal. Olds has advocated a change in attitude to view composting as a method of treatment with subsidy; the compost could be used for municipal purposes in parks, high way right-of-way, etc. One possible shortcoming found by Sturkie et al. was that soluble salts concentration could cause foliage damage.

Hydropulping

Because of the high costs associated with collection of wastes, new methods are constantly being tested as possible solutions. Hydro-pulping is a technique whereby the wastes, primarily paper and foods, are mascerated with water and transported to a central facility for dewatering and compassion of baling. This system is especially applicable in offices and sprawling establishments, such as hospitals or colleges. After the wastes are dewatered, they are removed from the premises and either landfilled or incinerated. Hence, hydropulping is a method of transportation and collection and not of disposal.

In office buildings hydropulping can result in minimizing the dusty conditions associated with solid waste collection, and it can reduce the size of janitorial staffs. The main advantage is the reduced cost and more efficient handing procedures.

There are some disadvantages associated with hydropulping; they include the need for large quantities of water to make the system operable. The water, while refuse is claimed, occasionally must be discharged to the sewer or treated to control odor and biological activity. The wet pulp which results after dewatering has a high moisture content which makes incineration difficult in conventional units.

Very little is known about the use of hydropulping as a collection and transportation technique. If some of the operation problems can be overcome, then the use of this method may prove to be economically competitive.

Pyrolysis

Pyrolysis is a thermal process where oxidation of the organic fraction is not allowed to occur. Instead, the organic matter is evolved from the refuse with heat,leaving an ash consisting mostly of tree carbon and any inorganic matter, e.g; metal and glass not removed before pyrolysis. Some of the gases which have been volatized are condensed while the remainder are burned to supply the heat (energy) needed to pyrolyze the material. Since oxidation is prevented, the pyrolysis process must be performed in an atmosphere of argon, helium or nitrogen.

Historically, pyrolysis has been used for a variety of purposes under the terminology of "destructive distillation." Coal has been pyrolyzed for the purpose of producing distillate products and coke. Commercial production of charcoal utilizes the principle of pyrolysis as the method for production. While the principle of pyrolysis is known and understood, the actual methanics of the system are less developed.

At the present time the potential applicability of pyrolysis to the disposal of solid waste is being investigated by the city of San Diego. All of the research has been on a pilot plants scale, but the results indicate that pyrolysis may be feasible. Scaling up the production to a full size plant may be the next step in utilizing the payrolysis technique.

Several pyrolysis processes are commercially available, and they have been used for production of charcoal, etc. While these methods have been successful for these materials, the applicability to solid waste pyrolysis is uncertain.

6 Management and Handling of Hazardous Substances

In exercise of the powers conferred by Sections 6,8 and 25 of the Environment (Protection) Act, 1986 (29 of 1986) and with a view to protecting the environment, nature and health, in connection with the application of gene technology and micro-organisms, the Central Government hereby makes the following rules, namely:-

1. Short title, extent and commencement

(1) The rules may be called the Rules for the Manufacture, Use, Import, Export and Storage of Hazardous micro-organisms/Genetically engineered organisms or cells.

(2) These rules shall come into operation on the date to be notified for this purpose in the Official Gazette.

2. Application

(1) These rules are applicable to the manufacture, import and storage of micro-organismsa nd Gene-technological products.

(2) These rules shall apply to genetically engineered organisms/micro-organisms and cells and correspondingly to any substances and products and food and stuffs, etc., of which such cells, organisms or tissues thereof form part.

(3) These rules shall also apply to new gene-technologies apart from those referred to in clauses

(ii) and (iv) of rule 3 and these rules shall apply to organisms/micro-organisms and cells generated by the utilisation of such other gene-technologies and to substances and products of which such organisms and cells form part.

(4) These rules shall be applicable in the following specific cases:

(a) sale, offers for sale, storage for the purpose of sale, offers and any kind of handling over with or without a consideration;

(b) exportation and importation of genetically engineered cells or organisms;

(c) production, manufacturing, processing, storage, import, drawing off, packaging and repackaging of the Genetically Engineered Products;

(d) production, manufacturing, etc. of drugs and pharmaceuticals and food stuffs, distilleries and tanneries, etc., which make use of micro-organisms/Genetically engineered micro-organisms one way or the other.

(5) These rules shall be applicable to the whole of India.

3. Definition

In these rules, unless the context requires,

(i) "Biotechnology" means the application of scientific and engineering principles to the processing of materials by biological agents to produce goods and services;

(ii) "Cell hybridisation" means the formation of live cells with new combinations of genetic material through the fusion of two or more cells by means of methods which do not occur naturally;

(iii) "Gene Technology" means the application of the gene technique called genetic engineering, include self cloning and deletion as well as cell hybridisation;

(iv) "Genetic engineering" means the technique by which heritable material which does not usually occur or will not occur naturally in the organism or cell concerned, generated outside the organism or the cell is inserted into said cell or organism. It shall also mean the formation of new combinations of genetic material by incorporation of a cell into a host cell, where they occur naturally (self cloning) as well as modification of an organism or in a cell by deletion and removal of parts of the heritable material;

(v) "Micro-organisms" shall include all the bacteria, viruses, fungi, mycoplasma, cell lines, algae, protodones and nematodes indicated in the schedule and those that have not been presently known to exist in the country.

4. Competent Authorities

(1) Recombinant DNA Advisory committee (RDAC). This committee shall review developments in Biotechnology at national and international levels and shall recommend suitable and appropriate safety regulations for India in recombinant research, use and applications from time to time. The Committee shall function in the Department of Biotechnology.

(2) Review Committee on Genetic Manipulation (RCGM): This committee shall function in the Department of Biotechnology to monitor the safety related aspects in respect of on-going research projects

and activities involving genetically engineered organisms/hazardous micro-organisms. The Review Committee on Genetic Manipulation shall include representatives of (a) Department of Biotechnology (b) Indian Council of Medical Research (c) Indian Council of Agricultural Research (d) Council of Scientific and Industrial Research (e) other experts in their individual capacity. Review Committee on Genetic Manipulation may appoint sub groups.

It shall bring out manuals of guidelines specifying procedure for regulatory process with respect to activities involving genetically engineered organisms in research, use and applications, including industry, with a view to ensure environmental safety. All on-going projects involving high risk category and controlled field experiments shall be reviewed to ensure that adequate precautions and containment conditions are followed as per the guidelines.

The Review Committee on Genetic Manipulation shall lay down procedures restricting or prohibiting production, sale, importation and use of such genetically engineered organisms or cells as are mentioned in the schedule.

(3) Institutional Biosafety Committee (IBSC): This committee shall be constituted by an occupier or any person, including research institutions, handling micro-organism/genetically engineered organisms. The Committee shall comprise the Head of the Institutions, scientists engaged in DNA work, a medical expert and a nominee of the Department of Biotechnology. The occupier or any person including research institutions handling micro-organisms/genetically engineered organisms shall prepare,with the assistance of the Institutional Biosafety Committee (IBSC) and up-to-date on-site emergency plan according to the manuals/

guidelines of the RCGM and make available copies to the District Level Committee/State Biotechnology Co-ordination Committee and the Genetic Engineering Approval Committee.

(4) Genetic Engineering Approval Committee (GEAC): This committee shall function as a body under the Department of Environment, Forests and Wildlife for approval of activities involving large scale use of hazardous micro-organisms and recombinants in research and industrial production from the environmental angle. The Committee shall also be responsible for approval of proposals relating to release of genetically engineered organisms and products into the environment, including experimental field trials.

The composition of the Committee shall be:

(i) Chairman—Additional Secretary, Department of Environment, Forests and Wild life.

Co-Chairman—Representative of Department of Biotechnology.

(ii) Members: Representatives of concerned Agencies and Departments, namely, Ministry of Industrial Development, Department of Ocean Development, Department of Science and Technology, Department of Biotechnology and the Department of Atomic Energy.

(iii) Expert members: Director General— Indian Council of Agricultural Research, Director General—Indian Council of Medical Research, Director General—Council of Scientific and Industrial Research, Director General—Health Services, Plant Protection Adviser, Directorate of Plant Protection, Quarantine and storage, Chairman, Central Pollution Control Board and three outside experts in individual capacity.

(iv) Member Secretary: An official of the Department of Environment, Forest and Wildlife.

The committee may co-opt other members experts as necessary.

The committee or any persons authorised by it shall have powers to take punitive action under the Environment (Protection) Act.

(5) State Biotechnology Co-ordination Committee (SBCC). There shall be a State Biotechnology Co-ordination Committee in the Sates wherever necessary. It shall have powers to inspect, investigate and take punitive action in case of violation of statutory provisions through the Nodal Department and the State Pollution Control Board/Directorate of Health/Medical Services. The Committee shall review periodically the safety and control measures in the various industries/institutions handling genetically engineered organisms/hazardous micro-organisms.

The composition of the Co—ordination Committee shall be:

(i) Chief Secretary - Chairman

(ii) Secretary, Department of Environment - Member Secretary

(iii) Secretary, Department of Health Member

(iv) Secretary, Department of Agriculture Member

(v) Secretary, Department

of Industries and Commerce - Member

(vi) Secretary, Department of Forests - Member

(vii) Secretary, Department of Public works/Chief Engineer, Department of Public Health Engineering - Member

(viii) State Microbiologists and Pathologists - Member

(ix) Chairman of State Pollution Control Board

The Committee may co-opt other members/experts as necessary.

(6) District Level Committee (DLC): There shall be a District Level Biotechnology Committee (DLC) in the districts wherever necessary under the District Collectors to monitor the safety regulations in installations engaged in the use of genetically modified organisms/hazardous micro-organisms and its applications in the environment.

The District Level Committee or any other persons authorised in this behalf shall visit the installation engaged in activity involving genetically engineered organisms, hazardous micro-organisms, formulate an information chart, find out hazards & risks associated with each of these installations & co-ordinate activities with a view to meeting any emergency. They shall also prepare an off-site emergency plan. The District Level Committee shall regularly submit its report to the State Biotechnology Co-ordination Committee and the Genetic Engineering Approval Committee. The District Level Committee shall comprise of:-

(i) District Collector - Chairman

(ii) Factory Inspector - Member

(iii) A representative of the Pollution Control Board - Member

(iv) Chief Medical Officer (District Health Officer) - Member (Convenor)

(v) District Agricultural Officer - Member

(vi) A representative of the Public Health Engineering Department - Member

(vii) District Microbiologists Pathologists (technical expert) - Member

(viii) Commissioner Municipal Corporation - Member

The Committee may co-opt other members/experts as necessary.

5. Classification of micro-organisms or genetically engineered product

(1) For the purpose of these rules, micro-organisms or genetically engineered organisms, products or cells shall be dealt with under two major heads; animal pathogens and plant pests and these shall be classified in the manner specified in the schedule.

(2) If any of the micro-organisms/genetically engineered organisms or cell falls within the limits of more than risk class as specified in the schedule, it shall be deemed to belong exclusively to the last in number of such classes.

6. Micro-organisms laid down in the schedule are divided into the following

(i) Bacterial Agents;

(ii) Fungal Agents;

(iii) Parasitic Agents;

(iv) Viral, Rickettsial and Chlamydial Agents;

(v) Special Category.

7. Approval and Prohibition, etc

(1) No person shall import, export, transport, manufacture, process, use or sell any hazardous micro-organisms or genetically engineered organism/substance or cell except with the approval of the Genetic Engineering Approval Committee.

(2) Use of pathogenic micro-organism or any genetically engineered organisms or cell for the purpose of research shall only be allowed in laboratories or inside laboratory areas notified by the Ministry of Environment and Forests for this purpose under the Environment (Protection) Act, 1986.

(3) The Genetic Engineering Approval Committee shall give directions to the occupier to determine or take measures concerning the discharge of micro-organisms/genetically engineered organisms or cells mentioned in the schedule from the laboratories, hospitals and other areas including prohibition of such discharges and laying down measures to be taken to prevent such discharges.

(4) Any person operating or using genetically engineered organisms/micro-organisms mentioned in the schedule for scale up or pilot operations shall have to obtain licence issued by the Genetic Engineering Approval Committee for any such activity. The possessor shall have to apply for licence in the prescribed proforma.

(5) Certain experiments for the purpose of education within the field of gene technology or micro-organisms may be carried outside the laboratories

and laboratory areas mentioned in sub-rule (2) and will be looked after by the Institutional Biosafety Committee.

8. Production

Production in which genetically engineered organisms or cells or micro-organisms are generated or used shall not be commenced except with the consent of Genetic Engineering Approval Committee with respect to the discharge of genetically engineered organisms or cells into the environment. This shall also apply to production taking place in connection with development, testing and experiments where such production etc. is not subject to rule 7.

9. Deliberate or unintentional release

(1) Deliberate or unintentional release of genetically engineered organisms/hazardous micro-organisms or cells, including deliberate release for the purpose of experiment shall not be allowed.

Note : Deliberate release shall mean any intentional transfer of genetically engineered organisms/ hazardous micro-organisms or cells to the environment or nature, irrespective of the way in which it is done.

(2) The Genetic Engineering Approval Committee may in special cases give approval of deliberate release.

10. Permission and approval for certain substances

Substances and products, which contain genetically engineered organisms or cells or micro-organisms shall not be produced, sold, imported or used except with the approval of the Genetic Engineering Approval Committee.

11. Permission and approval for food stuffs

Food stuffs, ingredients in food stuffs and additives,

including processing aids, containing or consisting of genetically engineered organisms or cells, shall not be produced, sold, imported or used except with the approval of the Genetic Engineering Approval committee.

12. Guidelines

(1) Any person who applies for approval under rules 8-11 shall, as determined by the Genetic Engineering Approval Committee submit information and make examinations or cause examinations to be made to elucidate the case, including examinations according to specific directions and at specific laboratories. He shall also make available an on-site emergency plan to the Genetic Engineering Approval Committee before obtaining the approval. If the authority itself makes an examination, it may order the applicant to defray the expenses incurred by it in doing so.

(2) Any person to whom an approval has been granted under rules 8-11 above shall notify the Genetic Engineering Approval Committee of any change in or addition to information already submitted.

13. Grant of approval

(1) In connection with the granting of approval under rules 8=11 above, terms and conditions shall be stipulated including terms and conditions as to the control to be exercised by the applicant, supervisions, restriction on use, the layout of the enterprise and as to the submission of information to the State Biotechnology Co-ordination Committee or to the District Level Committee.

(2) All approvals of the Genetic Engineering Approval Committee shall be for a specified period not exceeding four years at the first instance renewable for 2 years at a time. The Genetic Engineering Approval Committee shall have powers to revoke

such approval in the following situation:-

(a) if there is any new information as to the harmful effects of the genetically engineered organisms or cells;

(b) if the genetically engineered organisms or cells cause such damage to the environment, nature or health as could not be envisaged when the approval was given; or

(c) non-compliance of any condition stipulated by the Genetic Engineering Approval Committee.

14. Supervision

(1) The Genetic Engineering Approval Committee may supervise the implementation of the terms and conditions laid down in connection with the approvals accorded by it.

(2) The Genetic Engineering Approval Committee may carry out this supervision through the State Biotechnology Co-ordination Committee or the State Pollution Control Board/District Level Committee or through any person authorised in this behalf.

15. Penalties

(1) If an order is not complied with, the District Level Committee or State Biotechnology Co-ordination Committee may take measure at the expense of the person who is responsible.

(2) In case where immediate intervention is required in order to prevent any damage to the environment, nature or health the District Level Committee or State Biotechnology Co-ordination Committee may take the necessary steps without issuing any orders

or notice. The expenses incurred for this purpose will be repayable by the person responsible for such damage.

(3) The State Biotechnology Co-ordination Committee/ District Level Committee may take samples for a more detailed examination of organisms and cells.

(4) The State Biotechnology Co-ordination Committee/ District Level Committee shall be competent to ask for assistance from any other Government authority to carry out its instructions.

16. Responsibility to notify interruption or accidents

(1) Any person who under rules 7-11 is responsible for conditions or arrangements shall immediately notify the District Level Committee, State Bio-technology Co-ordination Committee and the state medical officer of any interruption of operations or accidents that may lead to discharges of genetically engineered organisms or cells which may be harmful to the environment, nature of health or involve any danger thereto.

(2) Any notice given under sub-rule (1) above shall not lessen the duty of the person who is responsible to try effectively to minimise or prevent the effects of interruption of operations or accidents.

17. Preparation of Off-site Emergency Plan by the DLC

(1) It shall be the duty of the DLC to prepare and off-site emergency plan detailing how emergencies relating to a possible major accident at a site will be dealt with and in preparing the plan, the DLC shall consult the occupier and such other person as in may deem necessary.

(2) For the purpose of enabling the DLC to prepare the emergency plan required under sub-rule (1), the

occupier shall provide the DLC with such information relating to the handling of hazardous micro-organisms/genetically engineered organisms under his control as the DLC may require, including the nature, extent and likely off-site effects of a possible major accident and the DLC shall provide the occupier with any information from the off-site emergency plan which relates to his duties under rule 16.

18. Inspection and information regarding finance

(1) The State Biotechnology Co-ordination Committee or the Genetic Engineering Approval Committee/the DLC or any person with special knowledge duly authorised by the State Biotechnology Co-ordination Committee or the Genetic Engineering Approval Committee or the DLC where it is deemed necessary, at any time on due production of identity be admitted to public as well as to private premises and localities for the purpose of carrying out supervision.

(2) Any person who is responsible for activities subject to rules 7-11 above shall at the request of District Level Committee or State Biotechnology Co-ordination Committee or the Genetic Engineering Approval Committee submit all such information, including information relating to financial conditions and accounts, as is essential to the authority's administration under these rules. He shall also allow supervision or inspection by the authorities or persons indicated in sub-rule (1).

(3) The Genetic Engineering Approval Committee may fix fees to cover, in whole or in part,the expenses incurred by the authorities in connection with approvals, examinations, supervision and control.

Appeal

(1) Any person aggrieved by a decision made by the Genetic Engineering approval Committee/State Biotechnology Co-ordination Committee in pursuance e of these rules may within thirty days from the date on which the decision is communicated to him, prefer an appeal to such authority as may be appointed by the Ministry of Environment and Forests provided that the Appellate Authority may entertain the appeal after the expiry of the said period of thirty days if such authority is satisfied that the appellant was prevented by sufficient cause from filing the appeal in time.

Exemption

The Ministry of Environment and Forests shall, wherever necessary, exempt occupier handling a particular micro-organism/genetically engineered organism from rules 7-11.

The manufacture, storage and import of hazardous chemicals rules, 1989

1. Short title and commencement

(1) These rules may be called the Manufacture, Storage & Import of Hazardous Chemicals Rules, 1989.

(2) They shall come into force on the date of their publication in the official Gazette.

2. *Definitions*

In these rules, unless the context otherwise requires,

(a) "Act" means the Environment (Protection) Act, 1986 (29 of 1986):

(b) "Authority" means an authority mentioned in column 2 of Sche-dule 5;

(c) "export" with its grammatical variations and

cognate expression, means taking out of India to a place outside India;

(d) "exporter" means any person under the jurisdiction of the exporting country and includes the exporting country, who exports hazardous chemicals;

(e) "hazardous chemical" means:-

(i) any chemical which satisfies any of the criteria laid down in Part I of Schedule 1 and is listed in Column 2 of Part II of this Schedule;

(ii) any chemical listed in Column 2 of Schedule 2;

(iii) any chemical listed in Column 2 of Schedule 3;

(f) "import" with its grammatical variation and cognate expression, means bringing into India from a place outside India;

(g) "importer" means an occupier or any person who imports hazardous chemicals;

(h) "industrial activity" means:-

 (i) an operations or process carried out in an industrial installation referred to in Schedule 4 involving or likely to involve one or more hazardous chemicals and includes on site storage or on-site transport which is associated with that operation or process, as the case may be; or

 (ii) Isolated storage;

 (iii) pipeline;

(i) "isolated storage" means storage of hazardous chemical, other than storage associated with an installation on the same site specified in Schedule 4 where that storage involves at least the qualities of that chemical set out in Schedule 2;

(j) "major accident" means an occurance including any

particular major emission, fire, or explosion involving one or more hazardous chemicals and resulting from uncontrolled developments in the course of an industrial activity or due to natural events leading to serious effects both immediate or delayed, inside or outside the installation, likely to cause substantial loss of life or property including adverse-effects on the environment;

(k) "pipeline" means a pipe (together with any apparatus and works asociated therewith) or system of pipes (together with any apparatus and works associated therewith) for the conveyance of a hazardous chemical other than a flammable gas as set out in Column 2 of Part II of Schedule 3 at a pressure of less than 8 bars absolute; the pipeline also includes inter-state pipelines;

(l) "Schedule" means Schedule appended to these rules;

(m) "site" means any location where hazardous chemicals are manufactured or processed, stored, handled, used, disposed of and includes the whole of an area under the control of an occupier and includes pipejetty or similar structure whether floating or not;

(n) "Threshold quantity" means:-

   (i) in the case of a hazardous chemical specified in Column 2 Schedule 2, the quantity of that chemical specified in the corresponding entry in Columns' 3 & 4;

   (ii) in the case of a hazardous chemical specified in Column 2 Part I of Schedule 3, the quantity of that chemical specified in the corresponding entry in columns 3 & 4 of that part;

   (iii) in the case of a substance of a class specified in Column 2 of Part II of

Schedule 3, the quantity of all substances of that class specified in corresponding entry in Columns 3 & 4 of that part.

3. Duties of authorities

Subject to the other provisions of these rules, the authorities shall perform duties as specified in Column 3 of Schedule 5.

4. General responsibility of the occupier during industrial activity

(1) The rules shall apply to,

(a) an industrial activity in which a hazardous chemical, which any of the criteria laid down in Part I Schedule 1 and is listed in Column 2 of Part II of this Schedule is or may be involved; and

(b) isolated storage in which is involved a threshold quantity hazardous chemical listed in Schedule 2 in Column 2 which is equal to or more than the threshold quantity specified in the Schedule for that chemical in Column 3 thereof.

(2) An occupier who has control of an industrial activity in terms of sub-rule (1) shall provide evidence to show that he has,

(a) indentified the major accident hazards; and

(b) taken adequate steps to -

(i) Prevent such major accidents and to limit their consequence to persons and the environment;

(ii) Provide to the persons working on the site with the information, training and equipment, including antidotes, necessary to ensure their safety.

5. Notification of major accident

(1) Where a major accident occurs on a site, the occupier shall forthwith notify the concerned authority as indentified in Schedule 5 of that accident and furnish thereafter to the concerned authortiy, a report relating to the accidents, instalments, if necessary, in Schedule 6.

(2) The concerned authority shall on receipt of the report, in accordance with sub-rule (1) of this rule, shall undertake a full analysis of the major accident and send the requisite information to the Ministry of Environment and Forests through an appropriate channel.

(3) Where an occupier has notified a major accident to the concerned authority under respective legislation, he shall be deemed to have complied with the requirements as per sub-rule run on (1) of this rule.

6. Industrial activity to which Rule 7 to 15 apply

(1) Rules 7 to 15 shall apply to;

(a) an industrial activity in which there is involved a quantity of a hazardous chemical listed in Column 2 of Schedule 3 which is equal to or more than the quantity specified in the entry for that chemical in Columns 3 & 4 (Rules 10-12 only for Column 4) and (b) isolated storage in which there is involved a quantity Specified in the entry for that chemical in Column 4.

(2) For the purposes of Rules 7 to 15,

(a) "new industrial activity" means an industrial activity which—

(i) commences after the date of coming into operation of these rules: or

(ii) if commenced before that date, is an industrial activity in which the modification has been made which is likely to cover major accident hazards, and that activity shall be deemed to have commenced on the date on which the modification was made;

(b) an "existing industrial activity" means an industrial activity which is not new industrial activity.

7. Notification of sites

(1) An occupier shall not undertake any industrial activity unless he has submitted a written report to the concerned authority containing the particulars specified in Schedule 7 at least 3 months before commencing that activity or before such shorter time as the concerned authority may agree and for the purposes of this paragraph, an activity in which subsequently there is or is liable to be a threshold quantity or more of an additional hazardous chemical shall be deemed to be a different activity and shall be notified accordingly.

(2) No report under sub-rule (1) needs to be submitted by the occupier if he submits a report under Rule 10 (1).

8. Updating of the site notification following changes in the threshold quantity

Where an activity has been reported in accordance with Rule 7 (1) and the occupier makes a change in it (including an increase or decrease in the maximum threshold quantity of a hazardous chemical to which this rule applies which is or is liable to be at the site or in the pipeline or at the cessation of the activity) which affects the particulars specified in that report or any subsequent report made under this rule, the occupier shall forthwith furnish a further report to the concerned authority.

9. Transitional provisions

Where —

(a) at the date of coming into operation of these rules, an occupier is in control of an existing industrial activity which is required to be reported under Rule 7 (1) or;

(b) within 6 months after that date an occupier commences any such new industrial activity; it shall be a sufficient compliance with that rule if he reports to the concerned authority as per the particulars in Schedule 7 within 3 months after the date of coming into operation of these rules or within 3 months after the date of coming into operation of these rules or within such longer time as the concerned authority may agree in writing.

10. Safety reports

(1) Subject to the following paragraphs of this rule, an occupier shall not undertake any industrial activity to which this rule applies, unless he has prepared a safety report on that industrial activity containing the information specified in Schedule 8 and has sent a copy of that report to that concerned authority at least 90 days before commencing that activity.

(2) In case of a new industrial activity which an occupier commences or by virtue of sub-rule (2) (a) (ii) of Rule 6 is deemed to commence, within 6 months after coming into operation of these rules, it shall be sufficient compliance with sub-rule (1) of the rule if the occupier sends to the concerned authority a copy of the report required in accodance with that sub-rule within 90 days after the date of coming into operation of these rules.

(3) In the case of existing industrial activity, until 5 years from the date of coming into operation of these rules, it shall be sufficient with sub-rule (1) of this rule if the occupier on or before 90 days from the date of coming into operation of these rules sends to the concerned authority the information specified in Schedule 7 relating to that activity.

11. Updating of reports under Rule 10

(1) Where an occupier has made a safety report in accordance with sub-rule (1) of Rule 10 he shall not make any modification to the industrial activity to which that safety report relates which could materially affect the particulars in that report unless he has made a further report to take account of those modifications and has sent a copy of that report to the concerned authority at least 90 days before making those modifications.

(2) Where an occupier has made a report in accordance with Rule 10 and sub-rule (1) of this rule and that industrial activity is continuing, the occupier shall within 3 years of the date of the last such report, make a further report which shall have regard in particular to new technical knowledge which has affected the particulars in the previous report relating to safety and hazard assessment, and shall within 30 days or in such longer time as the concerned authority may agree in writing, send a copy of the report to the concerned authority.

12. Requirements for further information to be sent to the authority

(1) Where in accordance with Rule 10, an occupier has sent a safety report relating to an industrial activity to the concerned authority, the concerned authority may, by a notice served on the occupier, require him to provide such additional information as is

specified in the notice and the occupier shall send that information to the concerned authority within such tme as is specified in the notice or within such extented time as the authority may subsequently specify.

13. Preparation of on-site emergency plan by the occupier

(1) An occupier shall prepare and keep up-to-date an on-site emergency plan detailing how major accidents will be date an on-site emergency plan detailing how major accidents will be dealt with on the site on which the industrial activity is carried on and that plan shall include the name of the person who is responsible for safety on the site and the names of those who are authorised to take action in accordance with the plan in case of an emergency.

(2) The occupier shall ensure that the emergency plan prepared in accordance with sub-rule (1), takes into account any modification made in the industrial activity and that every person on the site who is affected by the plan is informed of its relevant provisoins.

(3) the occupier shall prepare the emergency plan required under sub-rule (1),

   (a) in case of a new industrial activity, before that activity is commenced;

   (b) in the case of an existing industrial activity within 90 days of coming into operation of these rules.

14. Preparation of off-site emergency plans by the authority

(1) It shall be the duty of the concerned authority as identified in Column 2 of Schedule 5 to prepare and keep up-to-date an adequate off-site emergency plan detailing how mergencies relating to a possible

major accident on that site will be dealt with and in preparing that plan the concerned authority shall consult the occupier, and such other persons as it may deem necessary.

(2) For purpose of enabling the concerned authority to prepare the emergency plan required under sub-rule (1), the occupier shall provide the concerned authority with such information relating to the industrial activity under his control as the concerned authority may require, including the nature, extent and likely effects off-site of possible major accidents and the authority shall provide the occupier with any information from the off-site emergency plan which relates to his duties under Rule 11.

(3) The concerned authority shall prepare its emergency plan required under sub-rule (1)-

(a) in the case of a new industrial activity, before that activity is commenced;

(b) in the case of existing industrial activity, within 6 months of coming into operation of these rules.

15. Information to be given to presons liable to be affected by a major accident

(1) The occupier shall take appropriate steps to inform persons outside the site, either directly or through the district Emergency Authority, who are likely to be in an area which may be affected by a major accident about-

(a) the nature of the major accident hazard; and

(b) the safety measures and the 'dos' and 'don'ts' which should be adopted in the event of a major accident.

(2) The occupier shall take the steps required under subrule (1) to inform persons abut an industrial activity, before that activity is commenced, except, in the case of an existing industrial activity in which case the occupier shall comply with the requirements of sub-rule (1) within 90 days of coming into operation of these rules.

16. Disclosure of information

Where for the purpose of evaluating information notified under Rules 5 & 7 to 15, the concerned authority discloses that information to some other person, that other person shall not use that information for any purpose except for the purpose of the concerned authority disclosing it, and before disclosing the information the concerned authority shall inform that other person of his obligations under this paragraph.

17. Collection, development and dissemination of information

(1) This rule shall apply to an industrial activity in which a hazardous chemical which satisfies any of the criteria laid down in Part I of Schedule 1 and is listed in Column 2 of Part II of this Schedule is or may be involved.

(2) An occupier who has control of an industrial activity in terms of subrule (1) of this rule, shall arrange to obtain or develop informaton in the form of a safety data sheet as specified in Schedule 9. The information shall be accessible upon request for reference.

(3) The occupier while obtaining or developing a safety data sheet, as specified in Schedule 9 in respect of a hazardous chemical handled by him, shall ensure that the information is recorded accurately and reflects the scientific evidence used in making the hazard determination. In case, any significant information regarding hazard of a chemical is

available, it shall be added to the material safety data sheet as specified in Schedule 9 as soon as practicable.

(4) every container or a hazardous chemical shall be clearly labelled or marked to identify:

  (a) the contents of the container;

  (b) the name and address of the manufacturer or importer of the hazardous chemical;

  (c) the physical, chemical and toxicological data as per the criteria given in Part I of Schedule 1.

(5) In terms of sub-rule (4) of this rule, where it is impractical to label the chemical in view of the size of the container or the nature of the package, provision should be made for other effective means like tagging or accompanying documents.

18. Import of hazardous chemicals

(1) This rule shall apply to a chemical which satisfies any of the criteria laid down in Part I of Schedule 1 and is listed in Column 2 of Part II of the Schedule.

(2) Any person responsible for importing hazardous chemicals in India shall provide at the time of import or within 30 days for the date of import to the concerned authorities as identified in Column 2 of Schedule 5 the information pertaining to —

  (i) the name and address of the person receiving the consignment in India;

  (ii) the port of entry in India;

  (iii) mode of transport from the exporting country to India:

  (iv) the quantity of chemical(s) being imported; and

(v) complete product safety information.

(3) If the concerned authority at the State is satisfied that the chemical being imported is likely to cause a major accident, it may direct the importer to take such steps, including stoppage of such imports as the concerned authority may deem it appropriate.

(4) The concerned authority at the State shall simultaneously inform the concerned Port Authority to take appropriate steps regarding safe handling and storage of hazardous chemicals while off-loading the consignment within the port premises.

(5) Any person importing the hazardous chemicals shall maintain the records of the hazardous chemicals imported as specified in Schedule 10 and the records so maintained shall be open for inspection by the concerned authority at the State or the Ministry of Environment and Forests or an officer appointed by them in this behalf.

(6) The importer of the hazardous chemical or a preson working on his behalf shall ensure that transport of hazardous chemicals from port of entry to the ultimate destination is in accordance with the Central Motor Vehicles Rules, 1989, framed under the provisions of the Motor Vehicles Act, 1988.

19. Improvement notices

(1) If the concerned authority is of the opinion that a person has contravended the provisions of these rules, the concerned authority shall serve on him a notice (in this para referred to as "the imporvement notice") requiring that person to remedy the contravention as the case may be, the matters occasioning it within such period as may be specified in the notice.

(2) A notice served under sub-rule (1) shall clearly specify the measures to be taken by the occupier in remedying said contraventions.

20. *Power of the Central Government to modify the Schedules*

Central Government may, at any time, by notification in the Official Gazette make suitable changes in the Schedules.

Chemicals having the following values of acute toxicity and which, owing to their physical and chemical properties, are capable of producing major accident hazards.

Schedule I

(See Rules 2 (e), 4 (1) (a), 4(2), 17 & 18)

Indicative Criteria and List of Chemicals

Part I

(A) Toxic Chemicals

(B) Flammable chemicals

(i) flammable gases—chemicals which in the gaseous state at normal pressure and mixed with air become flammable and the boiling point which at normal pressure is 20°C or below;

(ii) highly flammable liquids— chemicals which have a flash point lower than 23°C and the boiling point of which at normal pressure is above 20°C.

(iii) flammable liquids—chemicals which have a flash point lower than conditions, such as high pressure and high temperature may create major accident hazards.

(C) Explosives

Chemicals which may explode under the effect of flame, heat or photochemical conditions or which are more sensitive to shocks or friction than Nitrobenzene.

Schedule 2

[See Rules 2(e) (ii), 4(1) (b), 4(2) and 6(1) (b)]

Isolated stroage at installations other than those covered by Schedule 4.

(A) The threshold quantities set out below relate to each installation or group of installations belonging to the same occupier where the distance between installations is not sufficient to avoid, inforceable circumstances, any aggravation of major accident hazards. These threshold quantities apply in any case to each group installations belonging to the same occupier where the distance between the installation is less than 500 metres.

(B) For the purpose of determining the threshold quantity of a hazardous chemical at an isolated storage, account shall also be taken of any hazardous chemical which is:-

(i) in that part of any pipeline under the control or the occupier having control of the site which is within 500 metres of that site connected to it;

(ii) at any other site under the control of the same occupier any part of the boundary of which is within 500 metres of the said site; and

(iii) in any vehicle, vessel, aircraft or hovercraft under the control of the same occupier which is used for storage purpose whether at the site within 500 meteres of it; but no account shall be taken of any hazardous chemical which is in a vehicle, vessel aircraft or hovercraft used for transporting it.

(a) This applies to ammonium nitrate & mixtures of ammonium nitrate where the nitrogen content derived from the ammonium nitrate is grater that 28% by weight and to aqueous solutions of ammonium nitrate where the concentration of ammonium nitrate is greater than 90% by weight.

(b) This applies to straight ammonium nitrate fertilizers and to compound fertilizers where the nitrogen content derived from the ammonium nitrate is greater than 28% by weight (a compound fertilizer contains ammonium nitrate together with phosphate and/or potash).

Shedule 3

[See Rules 2 (e) (iii), 5 and 6 (1) (a)]

List of Hazardous Chemical for application of rules 5 and 7 to 15.

(A) The quantities set-out below relate to each installation or a group of installations belonging to the same occupier where the distance between the installations is not sufficient to avoid, inforceable circumstances, any aggravation of major accident hazards. These quantities apply in any case to each group of installations belonging to the same occupier where the distance between the istallations is less than 500 metres.

(B) For the purpose of determining the threshold quantitiy of a hazardous chemical in an industrial installation, account shall also be taken of any hazardous chemical which is:-

(i) in that part of any pipeline under the control of the occupier having control of the site, which is within 500 metres of that site and connected to it;

(ii) at any other site under the control of the same occupier any part of the boundary of which is within 500 metres of the said site; and

(iii) in any vehicle, vessel, aircraft or hovercraft under the control of the same occupier which is used for storage purpose either at the site within 500 metre of it; but no account shall be taken of any hazardous chemical which is in a vehicle, vessel, air craft or hovercraft used for transporting it.

The Hazardous Wastes (Management and Handling) Rules, 1989

In exercise of the powers conferred by Sections 6, 8 and 25 of the Environment (Protection) Act, 1986 (29 of 1986) the Central Government bereby makes the following rules, namely :-

1. Short Title & Commencement

(1) These rules may be called the Hazardous Wastes (Management & Handling) Rules, 1989.

(2) They shall come into force on the date of their publication in the Official Gazette.

2. Application

These rules shall apply to hazardous wastes as specified in the Schedule and shall not apply to—

(a) Waste water and exhaust gases as covered under the provisions of the Water (Prevention and Control of Pollution) Act, 1974 (6 of 1974), and the Air (Prevention and Control of Pollution) Act, 1981 (14 of 1981), and rules made thereunder;

(b) wastes arising out of the operation from ships beyond five kilometers as covered under the provisions of the Merchant Shipping Act, 1958 (44 of 1958), and the rules made thereunder;

(c) radioactive wastes as covered under the provisions of the Atomic Energy Act, 1962 (33 of 1962), and rules made thereunder.

3. Definitions

In these rules unless the context otherwise requires :-

(a) 'Act' means the Environment (Protection) Act, 1986 (29 of 1986);

(b) 'applicant' means a person or an organisation that applies, in From 1, for granting of authorisation to perform specific activities connected with handling of hazardous wastes;

(c) 'authorisation' means permission for collection, reception, treatment, transport, storage and disposal of hazardous wastes, granted by the competent authority in From 2;

(d) 'authorised person' means a person or an organisation authorised by the competent authority to collect, treat, transport, store or dispose off hazardous wastes in accordance with the guidelines to be issued by the competent authority from time to time;

(e) 'export' with its grammatical variations and congnate expressions, means taking out of India to a place outside India;

(f) 'exporter' means any person under the jurisdiction of the exporting country who exports hazardous wastes and the exporting country itself, which exports hazardous wastes;

(g) 'facility' means any location wherein the process incidental to the waste generation, collection reception, treatment, storage and disposal are carried out'

(h) 'Form" means form appended to these rules;

(i) 'hazardous wastes' means categories of wastes specified in the Schedule;

(j) 'hazardous wastes site' means a place for collection, reception, treatment, storage and disposal of hazardous wastes which has been duly approved by the competent authority;

(k) 'import' with its grammatical variations and cognate expressions, means bringing into India from a place outside India;

(l) 'importer' means an occupier or any person who imports hazardous wastes;

(m) 'operator of a facility' means a person who owns or operates a facility for collection, reception, treatment, storage and disposal of hazardous wastes;

(n) 'Schedule' means the Schedule appended to these rules;

(o) 'State Pollution Control Board' means the Board appointed under sub-section (1) of Sec. 4 of the Water (Prevention and Control of Pollution) Act, 1974 (6 of 1974) and under Sec. 4 of the Air (Prevention and Control of Pollution) Act, 1981 (14 of 1981);

(p) 'transboundary movement' means any movement of hazardous wastes or other wastes from an area under the national jurisdiction of one country to or through an area under the national jurisdiction of another country of through an area not under the national jurisdiction of any other country, provided at least two countries are involved in the movement;

(q) the words and expressions used in these rules and not defined but defined in the Act, shall have the meanings respectively assigned to them in the Act.

4. *Responsibility of the occupier for handling of wastes*

(1) The occupier generating hazardous wastes listed in column (2) of the Schedule in quantities equal to or exceeding the limits given in column (3) of the said

Schedule, shall take all practical steps to ensure that such wastes are properly handled and disposed off without any adverse effects which may result from such wastes and the occupier shall also be responsible for proper collection, reception, treatment, storage and disposal of these wastes either himself or through the operator of a facility.

(2) The occupier or any other person acting on his behalf who intends to get his hazardous wastes treated by the operator of a facility under sub-rule (1) shall give to the operator of a facility such information as may be specified by the State Pollution Control Board.

5. *Grant of authorisation for handling hazardous wastes*

(1) Hazardous wastes shall be collected, treated, stored and disposed of only in such facilities as may be authorised for this purpose.

(2) Every occupier generating hazardous wastes and having a facility for collection, reception, treatment, transport, stroage and disposal of such wastes shall make an application in Form 1 to the State Pollution Control Board for the grant of authorisation for any of the above activities.

Provided that the occupier not having a facility for the collection, reception, treatment, transport, storage and disposal of hazardous wastes shall make an application to the State Pollution Control Board in Form 1 for the grant of authorisation within a period of six months from date of commencement of these rules.

(3) Any person, who intends to be an operator of a facility for the collection, reception, treatment, transport, storage and disposal of hazardous wastes, shall make an application in Form 1 to the State Pollution Control Board for the grant of authorisation for any of the above activities:

Provided that the operator engaged in the business of collection, reception, treatment, transport, storage and disposal of hazardous wastes shall make an application to the State Pollution Control Board in Form 1 for the grant of authorisation within a period of six months from the date of commencement of these ruels.

(4) The State Pollution Control Board shall not issue an authorisation unless it is satisfied that the operator of a facility or an occupier, as the case may be, possesses appropriate facilities, technical capabilities and equipment to handle hazardous wastes safely.

(5) The authorisation to operate a facility shall be issued in Form 2 and shall be sjubject to conditions laid down therein.

(6) (i) An authorisation granted under this rule shall, unless sooner suspended or cancelled, be in force a period of two years from the date of issue or from the date of renewal.

(ii) An application for the renewal of an authorisation shall be made in Form 1, before its expiry.

(iii) The authorisation shall continue to be in force until it is renewed or revoked.

(7) The State Pollution Control Board may, after giving reasonable opportunity of being heard to the applicant, refuse to grant any authorisation.

6. Power to suspend or cancel an authorisation

(1) The State Pollution Control Board may cancel an authorisation issued under these rules or suspend it for such period as it thinks fit if, in its opinion, the authorised person has failed to comply with any of the conditions of the authorisation or with any provisions of the Act or these rules, after giving the

authorised person an opportunity to show cause and after recording reasons therefor.

(2) Upon suspension or cancellation of the authorisation and during the pendency of an appeal under rule 12, the State Pollution Control Board may give directions to the person whose authorisation has been suspended or cancelled for the safe storage of the hazardous wastes, and such person shall comply with such directions.

7. Packaging, labelling and transport of hazardous wastes

(1) Before hazardous waste is delivered at the hazardous wastes site, the occupier or operator of a facility shall ensure that the hazardous waste is packaged in a manner suitable for storage and transport and the labelling and packaging shall be easily visible and be able to withstand physical conditions and climatic factors.

(2) Packaging, labelling and transport of hazardous wastes shall be in accordance with the provisions of the rules issued by the Central Government under the Motor Vehicles Act, 1988, and other guidelines issued from time to time.

8. Inventory of disposal sites

(1) The state Government or a person authorised by it shall undertake a continuing programme to identify the sites and compile and publish periodically an inventory of disposal sites within the State for the disposal of hazardous wastes.

(2) The State Government or a person authorised by it shall undertake an environmental impact study before identifying a site as waste disposal site in the State.

(3) The State Government or a person authorised by it

shall undertake a continuing programme to compile and publish an inventory of sites within the State at which hazardous wastes have at any time been stored or disposed of and such inventory shall contain, besides the location and description, information relating to the amount, nature and toxicity of hazardous wastes at each such site as may be associated with such site.

9. Records and returns

(1) The occupier generating hazardous waste and operator of a facility for collection, reception, treatment, transport, storage and disposal of hazardous wastes shall maintain records of such operations in Form 3.

(2) The occupier and operator of a facility shall send annual returns to the State Pollution Control Board in Form 4.

10. Accident reporting and follow-up

Where an accident occurs at the facility or on a hazardous waste site or during transportation of hazardous wastes, the occupier or operator of a facility shall report immediately to the State Pollution Control Board about the accident in Form 5.

11. Import of hazardous wastes

(1) Import of hazardous wastes from any country to India shall be permitted for dumping and disposal of such wastes. However, import of such wastes may be allowed for processing or reuse as raw material, after examining each case in merit by the State Pollution Control Board or by an officer authorised in this behalf.

(2) The exporting country or the exporter, as the case may be, of hazardous waste shall communicate in

Form 6 to the Central Government (the Ministry of Environment and Forests) of the proposed transboundary movement of hazardous wastes.

(3) The Central Government shall, after examining the communication received under sub-rule (2) and on being satisfied that the import of such hazardous wastes is to be used for processing or reuse as raw material, grant permission for the import of such wastes subject to such conditions as the Central Government may specify in this behalf and if, however, the Central Government is not satisfied with the communication received under sub-rule (2), it may refuse permission to import such hazardous wastes.

(4) Any importer importing hazardous wastes shall provide necessary information as to the type of hazardous wastes he is to import, in Form 6, to the concerned State Pollution Control Board/the Central Pollution Control Board in the case of Union Territories.

(5) The State Pollution Control Board shall examine the information received under sub-rule (4) and issue such instructions to the importer as it considers necessary.

(6) The Central Government or the State Pollution Control Board, as the case may be, shall inform the concerned Port Authority to take appropriate steps regarding the safe handling of the hazardous wastes at the time of off-loading the same.

(7) Any person importing hazardous wastes shall maintain the records of the hazardous wastes imported as specified in Form 7 and the records so maintained shall be open for inspection by the State Pollution Control Board/the Ministry of Environment and Forests/the Central Pollution

Control Boards in the case of Union Territories or an officer appointed by them in this behalf.

12. Appeal

(1) An appeal shall lie, against any order of suspension or cancellation or refusal of an authorisation by the State Pollution Control Board to the State Government and to the Ministry of Environment and Forests in the case of the Central Pollution Control Board.

(2) Every appeal shall be in writing and shall be accompanied by a copy of the order appealed against and shall be presented within thirty days of the order passed.

INDEX